ENERGY SCIENCE, ENGINEERING AND TECHNOLOGY

DIRECT METHANOL FUEL CELLS

APPLICATIONS, PERFORMANCE AND TECHNOLOGY

ENERGY SCIENCE, ENGINEERING AND TECHNOLOGY

Additional books in this series can be found on Nova's website under the Series tab.

Additional e-books in this series can be found on Nova's website under the eBooks tab.

ENERGY SCIENCE, ENGINEERING AND TECHNOLOGY

DIRECT METHANOL FUEL CELLS

APPLICATIONS, PERFORMANCE AND TECHNOLOGY

ROSE HERNANDEZ
AND
CARYL DUNNING
EDITORS

NOTICE TO THE READER

Library of Congress Cataloging-in-Publication Data

Names: Hernandez, Rose, editor. | Dunning, Caryl, editor.
Title: Direct methanol fuel cells : applications, performance and technology / editors, Rose Hernandez and Caryl Dunning.
Description: Hauppauge, New York : Nova Science Publishers, Inc., [2017] | Includes bibliographical references and index.
Identifiers: LCCN 2017044636 (print) | LCCN 2017046167 (ebook) | ISBN 9781536126044 | ISBN 9781536126037 (softcover) | ISBN 9781536126044 (ebook)
Subjects: LCSH: Direct methanol fuel cells. | Methanol as fuel.
Classification: LCC TK2933.D57 (ebook) | LCC TK2933.D57 D57 2017 (print) | DDC 621.31/2429--dc23
LC record available at https://lccn.loc.gov/2017044636

Published by Nova Science Publishers, Inc. † New York

CONTENTS

PREFACE

In Chapter One, L. Khotseng provide an overview of the recent advances in electrocatalysts for direct methanol fuel cells for both anode and cathode catalysts in order to present direct methanol fuel cells as an alternative power source for portable devices. In Chapter Two, Nobuyoshi Nakagawa, Mohammad Ali Abdelkareem, Takuya Tsujiguchi, and Mohd Shahbudin Masdar propose a fuel supply layer using a porous carbon plate (PCP) for DMFCs allowing for the use of high methanol concentrations. The proposed layer is comprised of a thin PCP layer, as well as a gap layer that has a mechanism to supply the methanol as a vapor. Next, Chapter Three by D.S. Falcão, J. P. Pereira, and A.M.F.R. Pinto review the multiphase flow in fuel cells modeling approaches while also reviewing the flow visualization techniques for flow analysis in fuel cells. Chapter Four by B.A. Braz, V.B. Oliveira, and A.M.F.R. Pinto closes the book by discussing the key work that has been done to improve the passive DMFC performance and providing a review on the most recent developments in passive DMFCs.

Chapter 1 - Direct methanol fuel cells (DMFCs) could be an alternative power source for portable devices. They are currently limited by the performance of their electrocatalysts due to methanol crossover from anode to cathode, which lowers the system's efficiency and causes CO poisoning and slow anode kinetics. Advanced electrocatalysts such as Pt alloys, and non-Pt catalysts, such as Pd, have enhanced the performance through improved structure, reduced particle size and improved composition. Catalyst supports have shown to enhance the activity of the catalysts through their surface area, dispersion and electronic and atomic structure effects. Various carbon-based supports, e.g., carbon nanotubes, graphene, carbon nanofibres and mesoporous carbon, have been used as supports for electrocatalysts for DMFCs. The performance of fuel cell catalysts is also based on their preparation methods; several preparation methods such as glycol, colloidal, impregnation and micro-emulsion have been developed with improved structural strength, reduced loading and better activity at a higher efficiency. This chapter aims to cover recent advances in electrocatalysts for DMFCs for both anode and cathode catalysts, their supports and their preparation methods.

Chapter 2 - In direct methanol fuel cells, methanol crossover (MCO) through the electrolyte membrane is a major problem to be solved. Due to the MCO, degradation of the power generation and energy loss occur, hence a diluted methanol solution with 1 to 3 M (3 to 10 wt.%) is generally used as the fuel, which limits achieving a high energy density that is an attractive feature of the DMFC. The authors have proposed a fuel supply layer using a porous carbon plate (PCP) for DMFCs that allowed using high methanol concentrations up to neat methanol. The layer consists of a thin PCP and a gap layer that has a mechanism to supply the methanol as a vapor. This effective and practical technique is suitable for the utilization of a high methanol concentration and miniaturization of the DMFC system. In this chapter, the authors described the principle and practical MCO control at the fuel supply layer using the PCP. The power generation performance of the DMFCs equipped with the fuel supply layer, the mass transport characteristics through the membrane, and the development of prototype stacks are explained. The pore structure and wettability of the PCP were found to have a significant impact on the MCO. Since the fuel supply is conducted in the vapor phase, water transport through the membrane was quite different from that by the liquid feed, as well as affecting the cathode performance and reaction products. For practical applications, prototypes of the DMFC's stack with the PCP were designed and tested. The power generation profiles of the passive and active stacks are then discussed in detail in this chapter.

Chapter 3 - Two-phase flow phenomena occur in low temperature direct methanol fuel cells through the generation of CO_2 gaseous bubbles at the anode side, or due to the accumulation of water droplets at the cathode side. In order to achieve better cell performances, two-phase flow characterization is crucial when designing these devices. Therefore, flow visualization is fundamental to better understand the effect of two-phase flows at macro- and especially at micro-scales. Particle Image Velocimetry (PIV) is a measurement technique that can be applied to both scales. This technique has received increased attention over the last two decades, since it allows the experimental validation of the multiphase behavior simulations usually performed by Computational Fluid Dynamics (CFD). In this chapter, a survey of the modeling approaches concerning multiphase flow in fuel cells is carried out. An attempt is made to review the flow visualization techniques applied for flow analysis in fuel cells, and a short description of the PIV methodology is provided, considering both macro and micro-scales. Additionally, the applications of this technique for the visualization and characterization of two-phase flow in fuel cells are dealt in detail. The advances made in this area are considered of increasing interest, since this technique may provide insight for the improved design and performance of fuel cells at the micro-scale.

Chapter 4 - Direct methanol fuel cells (DMFCs) are in the Energy Agenda due to their market potential to replace the conventional batteries in portable applications, either for recreational, professional, military and medical purposes, since they enable the direct conversion of the chemical energy stored in a fuel, methanol, to electrical energy with water and carbon dioxide as final products. Moreover, DMFCs offer higher energy densities than batteries, longer runtime, instant recharging, use a liquid fuel, operate at room temperature and can rely on a passive flow of reactants, allowing to operate the cell without any additional power consumption. Therefore, in the last years, the passive DMFCs systems, where the fuel and the oxidant are feed in a passive mode, have been studied and some progress towards its commercialization has already been achieved. However, the current systems still have higher costs, lower durability and power outputs due to the slow electrochemical reactions, water and methanol crossover and an inefficient products removal from the anode (gaseous carbon dioxide) and cathode sides (liquid water), which may hinder the fuel and the oxidant supply to the reaction zone. This chapter discusses the key work done in order to improve the passive DMFC performance based on the challenges of this technology. The main goal is to provide a review on the most recent developments in passive DMFCs and on the recent work concerning the optimization of the operating and design conditions and on empirical and fundamental modelling. This chapter starts by a brief introduction recalling the passive DMFCs technology followed by a description of the fundamentals of these systems. Then an intensive review on the recent experimental and modelling works performed with these cells is presented. Towards the introduction of passive DMFCs in the market, studies regarding its lifetime and durability, as well as a cost benefit analysis where this technology is compared with the traditional ones is also presented.

In: Direct Methanol Fuel Cells
Editors: Rose Hernandez et al.
ISBN: 978-1-53612-603-7

Chapter 1

ELECTROCATALYSTS FOR DIRECT METHANOL FUEL CELLS

L. Khotseng[*]
Department of Chemistry, University of the Western Cape,
Cape Town, South Africa

ABSTRACT

Direct methanol fuel cells (DMFCs) could be an alternative power source for portable devices. They are currently limited by the performance of their electrocatalysts due to methanol crossover from anode to cathode, which lowers the system's efficiency and causes CO poisoning and slow anode kinetics. Advanced electrocatalysts such as Pt alloys, and non-Pt catalysts, such as Pd, have enhanced the performance through improved structure, reduced particle size and improved composition. Catalyst supports have shown to enhance the activity of the catalysts through their surface area, dispersion and electronic and atomic structure effects. Various carbon-based supports, e.g., carbon nanotubes, graphene, carbon nanofibres and mesoporous carbon, have been used as supports for electrocatalysts for DMFCs. The performance of fuel cell catalysts is also based on their preparation methods; several preparation

---

[*] Corresponding Author Email: lkhotseng@uwc.ac.za.

methods such as glycol, colloidal, impregnation and micro-emulsion have been developed with improved structural strength, reduced loading and better activity at a higher efficiency. This chapter aims to cover recent advances in electrocatalysts for DMFCs for both anode and cathode catalysts, their supports and their preparation methods.

Keywords: anode catalyst, cathode catalyst, carbon black, mesoporous carbon, nanostructured carbon, impregnation, colloidal, micro-emulsion

INTRODUCTION

Electrocatalysts play an important role in the development of direct methanol fuel cells (DMFCs), as it is one of the key components of fuel cells. Significant progress has been made in the development of DFMC. However, there are still many challenges in this technology, including high costs, low activity and poor durability and reliability, which further discourage the application of DMFC, hence its unsuccessful commercialisation. These issues are closely related to electrocatalysts. DMFC requires platinum (Pt) based catalysts for the conversion of fuel at the anode and reduction of oxygen at the cathode; the high costs of Pt accounts for nearly half the cost of a fuel cell [1]. This chapter focuses on recent progress in the development of noble and non-noble electrocatalysts for anode and cathode electrocatalysts and their supports, which is a short-term solution in reducing costs, and on approaches employed to improve activity, reliability and durability for both methanol oxidation and oxygen reduction reactions (ORRs).

ANODE CATALYSTS FOR DMFC

Among challenges in terms of the development of electrocatalysts for DMFCs, anode catalysts suffer from high costs, slow reaction kinetics due to lower temperatures of DMFC and methanol crossover, which causes degradation and CO poisoning, which in turn strongly adsorbs the electrocatalysts, reducing the surface area and hence the performance of the DMFC [2]. These challenges have advanced the development of new improved electrocatalysts.

Very few electrode materials are capable of methanol adsorption; in acid, only Pt and Pt-based catalysts have been found to show both sensible activity and stability and are the most practical catalysts for DMFC [3, 4]. For pure Pt, the adsorption of CO, one of the intermediates in methanol oxidation, can occupy the reaction-active sites, causing slow reaction kinetics [5, 6].

The first complete reaction mechanism of Pt was formulated by Bagotzky et al. [7]. The latest review of methanol oxidation on platinum catalyst was developed by Leger [8]. Besides the reactions proposed by Bagotzky, Leger introduced the first hydrogen atom from the OH group in methanol in the mechanism [8, 9]. The mechanism of methanol oxidation can be summarised in terms of two basic functionalities:

1. Electrosorption of methanol onto the substrate
2. Addition of oxygen to adsorbed carbon-containing intermediates to generate CO_2.

$$\text{Pt-}(CH_3OH)_{ads} \rightarrow \text{Pt-}(CH_3O\bullet)_{ads} + H^+ + e^- \quad (1a)$$

Further transformation of formed species is enabled through dissociation of H atoms bound to a C atom in the following reactions:

$$\text{Pt-}(CH_3O\bullet)_{ads} \rightarrow \text{Pt-}(CH_2O\bullet)_{ads} + H^+ + e^- \quad (2a)$$

$$\text{Pt-}(CH_2O\bullet)_{ads} \rightarrow \text{Pt-}(CHO\bullet)_{ads} + H^+ + e^- \quad (3a)$$

CHO can also be formed in the following reactions:

$$\text{Pt-}(CH_3OH)_{ads} \rightarrow \text{Pt-}(\bullet CH_2OH)_{ads} + H^+ + e^- \quad (1b)$$

$$\text{Pt-}(\bullet\text{CH}_2\text{OH})_{ads} \rightarrow \text{Pt-}(\bullet\text{CHOH})_{ads} + \text{H}^+ + \text{e-} \quad (2b)$$

$$\text{Pt-}(\bullet\text{CHOH})_{ads} \rightarrow \text{Pt-}(\bullet\text{CHO})_{ads} + \text{H}^+ + \text{e-} \quad (3b)$$

• CHO is considered to be an active intermediate and can act further giving adsorbed CO, which is considered to be a poisoning intermediate on Pt,

$$\text{Pt-}(\bullet\text{CHO})_{ads} \rightarrow \text{Pt-}(\bullet\text{CO})_{ads} + \text{H}^+ + \text{e-} \quad (4a)$$

or can directly transform into CO_2,

$$\text{Pt-}(\bullet\text{CHO})_{ads} + \text{Pt-}(\text{OH})_{ads} \rightarrow 2\text{Pt} + \text{CO}_2 + 2\text{H}^+ + 2\text{e-} \quad (4b)$$

or first react in the following manner:

$$\text{Pt-}(\bullet\text{CHO})_{ads} + \text{Pt-}(\text{OH})_{ads} \rightarrow \text{Pt} + \text{Pt-}(\bullet\text{COOH})_{ads} + \text{H}^+ + \text{e-} \quad (5)$$

and then undergoes conversion to CO_2:

$$\text{Pt-}(\bullet\text{COOH})_{ads} \rightarrow \text{Pt} + \text{CO}_2 + \text{H}^+ + \text{e-} \quad (6)$$

At more positive potentials adsorbed CO can be oxidised through the following reactions:

$$\text{Pt-}(\bullet\text{CO})_{ads} + \text{Pt-}(\text{OH})_{ads} \rightarrow \text{Pt-}(\bullet\text{COOH}) \quad (7)$$

$$\text{Pt-}(\bullet\text{COOH})_{ads} \rightarrow \text{Pt} + \text{CO}_2 + \text{H}^+ + \text{e-} \quad (8)$$

Reactions (1) to (4) are electrosorption processes, whereas following reactions involve oxygen transfer or oxidation of surface-bonded intermediates. The formation of OH on the Pt surface is the necessary step for oxidative removal of adsorbed CO, which requires a high potential. In terms of methanol oxidation, such a high potential limits the fuel cell application of a pure Pt catalyst.

The oxidation kinetics can be improved significantly, even reaching a practicable level [10] by alloying the Pt with a second and sometimes with a third metal. The role of the second alloying metal is to improve the electronic properties of Pt by donating the d electron density. A second metal is able to provide oxygenated species at lower potentials, for oxidative removal of adsorbed CO is definitely needed [11]. It has also been noted that when the Pt is alloyed with oxophilic metals, in the bifunctional model, oxophilic metals are known to provide sites for water adsorption. It is reported that Beden et al. came up with the complete mechanism of the adsorption of methanol at Pt electrodes [12, 13].

Methanol dissociative adsorption:

$$Pt\text{-}(\bullet CHO)_{ads} \rightarrow Pt\text{-}(\bullet CO)_{ads} + H^+ + e^- \quad (4a)$$

Water dissociative adsorption:

$$Ru + H_2O \rightarrow Ru(OH)_{ads} + H^+ + e^- \quad (9)$$

CO electro-oxidation:

$$PtCO_{ads} + Ru(H_2O)_{ads} \rightarrow CO_2 + Pt + Ru + 2H^+ + 2e^- \quad (10)$$

The bifunctional mechanism involves the adsorption of oxygen-containing species on Ru at lower potentials, thereby promoting the oxidation of CO to CO_2. Reactions (9) and (10) are the rate-determining steps.

Binary Pt alloys, such as PtSn, PtRu, PtOs, PtW, PtMo, PtNi, PtMn, PtCo and PtW, have been investigated in order to improve methanol oxidation. Binary metals such as PtNi, PtMn, PtCo, PtMo and PtW [14, 15, 16] showed improved electrocatalytic activity compared to Pt [17] because of the change in the electronic structure of Pt by electron transfer of the co-catalyst to Pt [18]. The improved electrocatalytic activity is attributed to these metals being oxiphilic, hence behaving like a catalyst promoter. Among the various types of binary catalysts, PtRu alloys have shown reasonable activity and stability and are the most practical electrocatalysts for DMFC [19, 20]. The improved activity of PtRu catalysts when compared to Pt for methanol oxidation can be explained by bifunctional catalysis, where both the electronic effect (Ru promotes adsorption of methanol at lower potentials by providing a number of kinetically more active sites for methanol chemisorption) and the ability of Ru to be oxidised at lower overpotentials than Pt itself are arguments to understand the role of Ru as promoter [5, 8, 14, 16].

Although catalysts such as PtRu improve activity of the DMFC, the fact that both Pt and Ru are noble metals and also a high catalyst loading is required (2-8 mg/cm^2) in order to get an acceptable performance of a fuel cell, PtRu needs to be reduced through further improvement of activity and optimisation of catalyst in order to reduce the costs, which is hindering the commercialisation of fuel cells.

The above mentioned problem can be solved by modifying the binary catalyst with another metal or metal oxide that has higher tendencies to form surface-oxygenated species at lower potentials.

More work has been done to synthesise new catalysts for methanol oxidation based on PtRu ternary catalysts such as PtRuNi, PtRuMo and metal oxides. Researchers such as Wang et al. [21] and Gotz et al. [16] reported on methanol oxidation of various ternary PtRu-based catalysts, including PtRuNi/C (Vulcan XC-72), PtRuSn/C, PtRuW/C and PtRuMo/C as compared to PtRu/C. Their work using electrochemical techniques showed high electrocatalytic activity of PtRuNi/C PtRuW/C and PtRuMo/C compared to PtRu/C, while another ternary PtRu, i.e., PtRuSn, was less active. The explanation for higher electrocatalytic activity of those metals is that their promotion effect is ascribed to the bifunctional mechanism and electronic effect of Mo metal. According to the bifunctional effect, Mo mixed with Ru further supports H_2O activation, in comparison to Ru alone, which subsequently facilitates CO oxidation, resulting in enhanced MOR activity. Figure 1 presents a chart of different PtRu-based ternary catalysts published by Lamy et al. [22].

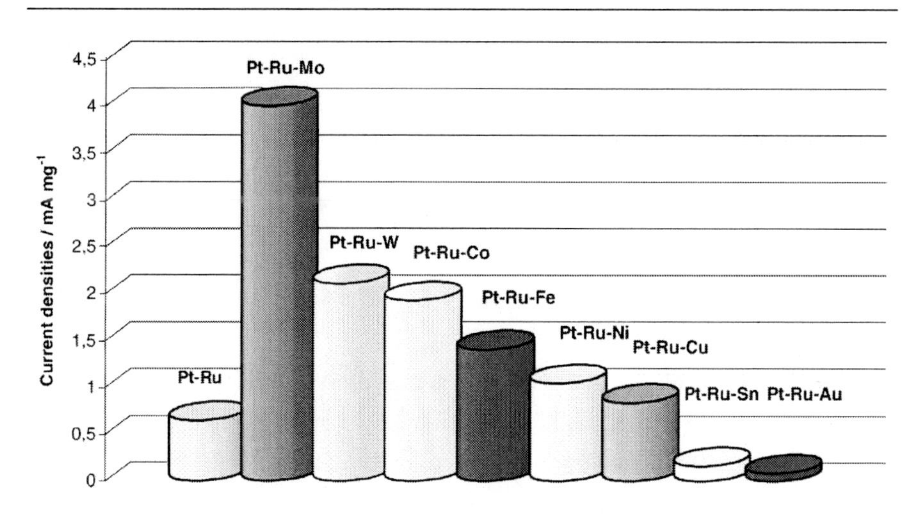

Figure 1. Oxidation of 1.0 M methanol in 0.1 M perchloric acid at room temperature at different Pt ternary electrocatalysts dispersed in a polyaniline film. Current taken after 30 min. at 450 mV vs. RHE [22]. [Reprinted with permission from Elsevier]

There are also reports on the improvement of CO tolerance and electrocatalytic activity of methanol oxidation with the addition of certain kinds of metal oxides, such as MoO_3, WO_3, TiO_2, etc. Ioroi et al. [23] investigated alloying MoO_x with Pt. From their findings, the CO tolerance for PtMoOx is higher as compared to PtRu/C, mainly due to the higher electrocatalytic activity of $PtMoO_x$/C catalysts as compared to PtRu/C and of Pt/C. They concluded that the alleviation of CO poisoning for $PtMoO_x$/C was due to the interaction between Pt and MoO_x such as an activation of the interface of Pt and MoOx against CO oxidation and the modification of the electronic band structure.

Other noble metals, such as Au, have also been synthesised as anode catalysts for DMFC. Lu et al. [24] synthesised ultrathin Au nanowires and studied their methanol oxidation properties in acidic and alkaline media. The Au nanowires exhibited higher electrocatalytic activity for methanol than the poly-Au nanoparticles and bulk Au in both acidic and alkaline media. Zhang et al. [25] fabricated nanostructured PtAu catalysts. These porous PtAu catalysts have better electrocatalytic activity, anti-poisoning ability and long-term structural stability than commercial PtRu catalysts.

In general, the MOR on an Au electrode proceeds in two distinct potential regions, with different mechanisms involving the two equations shown below:

$$CH_3OH + 5OH^- \rightarrow HCOO^- + 4H_2O + 4e^- \quad (11a)$$

$$CH_3OH + 8OH^- \rightarrow CO_3^{2-} + 6H_2O + 6e^- \quad (11b)$$

At lower potentials, the methanol is mainly oxidised to formate ion via an overall four-electron transfer reaction (eqn. 11a). At higher potentials, the methanol molecules were oxidised to carbonates with the exchange of a six-electron transfer reaction (eqn. 11b).

The nature, structure and composition of co-catalysts have an important effect on methanol oxidation in terms of activity and selectivity [26]. Recently, with the emergence of alkaline membranes that conduct hydroxide ions (OH-), alkaline direct alcohol fuel cells have attracted interest. The most significant advantage associated with the change in the electrolyte membrane from acid to base is that the reaction kinetics of both the alcohol anodic reaction and the oxygen cathodic reaction in alkaline media become faster than in acidic media, thus making it possible to use non-noble metal catalysts to reduce metal loading.

The mechanism of methanol in alkaline solution proceeds differently than that of Pt electrodes in the following steps [7, 22, 27]:

$$M + OH^- \rightarrow M\text{-}(OH)_{ads} + e^- \quad (12)$$

$$M + (CH_3OH)_{sol} \rightarrow M\text{-}(CH_3OH)_{ads} \quad (13)$$

$$M\text{-}(CH_3OH)_{ads} + OH^- \rightarrow M\text{-}(CH_3O)_{ads} + H_2O + e^- \quad (14)$$

$$M\text{-}(CH_3O)_{ads} + OH^- \rightarrow M\text{-}(CH_2O)_{ads} + H_2O + e^- \quad (15)$$

$$M\text{-}(CH_2O)_{ads} + OH^- \rightarrow M\text{-}(CHO)_{ads} + H_2O + e^- \quad (16a)$$

$$M\text{-}(CHO)_{ads} + OH^- \rightarrow M\text{-}(CO)_{ads} + H_2O + e^- \quad (17)$$

$$M\text{-}(CO)_{ads} + OH^- \rightarrow M\text{-}(COOH)_{ads} + H_2O + e^- \quad (18)$$

$$M\text{-}(COOH)_{ads} + OH^- \rightarrow M + CO2 + H_2O + e^- \quad (19)$$

where M stand for Pt or Pd.

In the presence of adequate alkali, the oxidation of M-(CHO) $_{ads}$ and M-(CO) $_{ads}$ may proceed directly through steps (17) and (18):

$$M\text{-}(CHO)_{ads} + 3OH^- \rightarrow M + CO_2 + 3e^- \quad (16b)$$

$$M\text{-}(CO)_{ads} + 2OH^- \rightarrow M + CO_2 + 2e^- \quad (16c)$$

The strength of the bonding of $(CHO)_{ads}$ on the surface determines the entire rate of the reaction. In general, the chemisorbed bonding of (CHO) $_{ads}$ on Pt metal catalysts in alkali is weak, such that further oxidation occurs without difficulty, that is, without irreversibly blocking the electrode's active sites. The complete removal of CO might not be possible at lower potentials, as they strongly bind to the electrode surface. However, at higher potential, highly active oxygen atoms again become available and the poisonous PtCO or Pd_2CO species is oxidised via the reaction (17) or (16c).

Pd, which belongs in the same group with Pt in the periodic table, hence have similar properties, the same fcc structure and the same atomic size, has also been investigated as the replacement for Pt. Pd as electrocatalyst for methanol oxidation reaction (MOR) in alkaline media has been proven to enhance higher activity. Pd compared to Pt is 50% more abundant on earth and the cost is lower than Pt [11, 28]. Pd has gained considerable attention due to its remarkable advantages, such as large surface area and high stability [29]. Pd has been used to oxidise small organic molecules, including methanol. Pd has the power to reduce protons, store and release hydrogen and is further used to remove adsorbed CO formed from methanol electroxidation, thereby decreasing the poisonous effect. It has been reported that Pd can assist in the oxidation of alcohol to CO_2 through the reaction of CO with hydrogen adsorbed in the Pd lattice at remarkably negative potentials [8]. The release of hydrogen by Pd thus provides a feasible route for lowering the surface concentration of adsorbed CO, permitting the continual oxidation of organic molecules at the Pd surface.

Pd is completely inactive as the electro-oxidation of methanol in acid media, while Takamura et al. [30] found that Pd exhibited higher electrocatalytic activity for methanol oxidation and better steady-state behaviour in alkaline solutions compared to Pt. They further studied the kinetics of MOR and concluded that the dehydrogenation of methanol occurred faster and that the rate-determining step was oxidative removal of methoxi radicals by the hydroxide ions [30]. Pd-based catalysts, including Pd oxide composites, Pd-based binary and ternary catalysts, have been tested by several researchers to improve activity and stability.

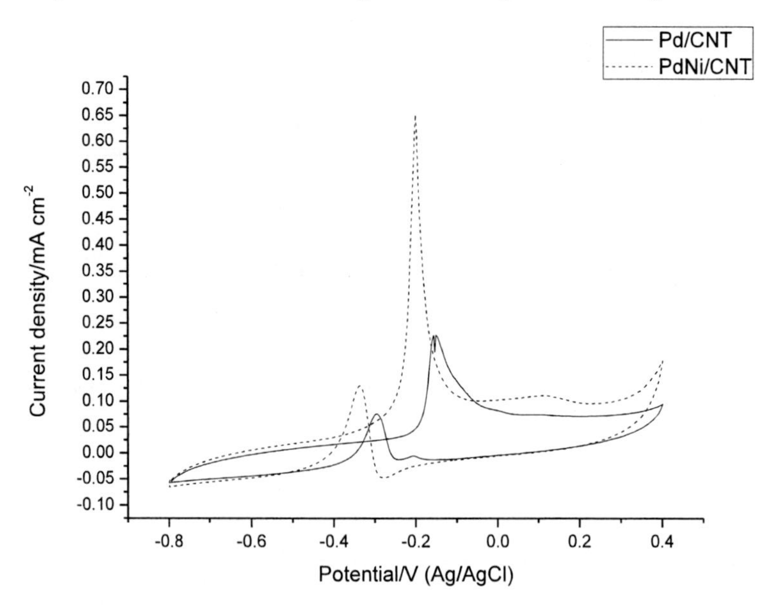

Figure 2. The cyclic voltammetry curve of methanol electro-oxidation on Pd/CNT (carbon nanotube) and PdNi/CNT catalyst with a solution of 1 M methanol and 1 M KOH at 25◦C. Scan rate: 30 mVs^{-1}.

Pd has been tested in fuel cells as the Pt co-catalyst in acid media. In an alkaline solution, the exchange current density of PtPd on MOR showed the maximum for a surface composition of approximately 15% in Pd [28]. This combined effect is relatively important, as exchange current densities are up to 10 times greater than Pt. Conversely, in acid the exchange current densities decreased from pure Pt to pure Pd. The combination of PtPd in an acid solution, which is inactive, results in diluting the Pt sites, which prevents the presence of three adjacent sites necessary for the formation of strongly bound intermediates as well as in the surface composition of > 33% in Pd.

Mahapatra et al. [8] compared Pd/C, Pt/C and PdPt/C in an alkaline medium. The electrocatalytic activity of PdPt/C was 15 and 5.8 times higher than Pt/C and Pd/C, respectively. This high value was attributed to the smaller particle size and high particle dispersion of PdPt/C. The electro-oxidation of three catalysts was compared and PdPt showed lower onset potential and higher current density compared to Pd/C and Pt/C. Hence, PdPt/C demonstrates higher electrocatalytic activity towards methanol oxidation, which is attributed to the fact that the alloying of Pd with Pt is capable of controlling the poisoning effect. Ru promoted the electrocatalytic activity of Pt towards methanol oxidation by providing oxygen-containing species at more negative potentials. Chen et al. [29] showed that the incorporation of Ru into Pd improved methanol oxidation kinetics in alkaline media. They found that the electrocatalytic activity of PdRu was much higher than that of Pd towards methanol oxidation. The incorporation of Ru gave rise to the negative shift of 0.15 V in the onset potential, with PdRu 1: 1 atomic ratio giving the highest potential. However, when PdRu was compared with PtRu for methanol oxidation in alkaline media, the activity of PdRu was higher than Pd, but lower than PtRu.

Jurzinsky et al. and Qi et al. [33, 34] compared PdNi/C to Pd/C. The results showed low onset potential for PdNi/C compared to Pd/C. In addition, PdNi/C obtained higher mass current densities and lower onset potential than Pd/C, which indicates more efficient electrocatalytic activity. The CVs of methanol electrooxidation on Pd/CNT (carbon nanotube), and

PdNi/CNT electrodes are displayed in Fig. 2. The PdNi/CNT cyclic voltammogram showed that the potential beginning of the methanol electro-oxidation (onset potential) was lower than Pt/NiCNT (-0.45V vs. Ag/AgCl); compared to -0.25V for Pt/CNT with higher current density towards methanol electro-oxidation in alkaline medium in the whole potential range evaluated, confirming that PtNi/CNT binary electrocatalysts are more electro-active compared to Pt/CNT electrocatalysts. This enhanced activity was attributed to the following reasons: The electronegativity of Pd (2.20) is larger than of Ni (1.91), thus the transfer of electrons from Ni to Pd may occur, which can decrease the Pd-CO binding energy, improve CO oxidation from methanol dehydrogenation, and enhance the adsorption and oxidation of methanol molecules. In addition, like Ru, Ni is an oxophilic element and has the capacity to generate OH_{ads} at a lower potential and facilitates the oxidative desorption of the intermediate products, thereby enhancing both the electrocatalytic activity and the stability of Pd catalysts [35].

The addition of metal oxides such as CeO_2, NiO, Co_3O_4 and Mn_3O_4 significantly promoted electrocatalytic activity and stability of the Pd/C catalysts for methanol oxidation, with $PdCo_3O_4$ showing a higher activity for methanol [26, 33, 36] in an alkaline medium than the Pt/C catalyst. The improvement is attributed to the fact that the OH_{ads} could easily be formed on metal oxides. OH_{ads} functions in the same manner as Ru in a PtRu catalyst, thereby enhancing the anti-poisoning ability. It is however difficult to control the size and morphology of metal oxides, which results in poor contact between the metal catalyst and the carbon powder and thus sacrifices some of the active sites of the prepared catalysts.

CATHODE CATALYSTS FOR DMFC

For cathode catalysts, Pt has the highest electrocatalytic activity for oxygen reduction of any of the pure metals, and when supported on the conductive carbon [28]. For DMFCs, due to the major problem of methanol crossover through polymer electrolyte, methanol adsorbs the Pt sites in the cathode for the direct reaction between methanol and oxygen. Hence, high loadings of expensive Pt catalyst are required from slow oxygen reduction reaction (ORR) due to mixed potential that occurs between oxygen reduction and methanol oxidation simultaneously, hence causing overpotential losses that amount to 0.3 to 0.4 V [37-40].

The development of more active oxygen reduction catalysts than Pt has been the subject of extensive research for a number of decades. The search for ORR catalysts that are more active, less expensive and have greater stability than Pt has led to the development of a Pt alloy and Pt-free catalysts, shape control of Pt nanocrystals and Pt with novel nanostructures [41, 42].

The electrocatalytic activity of ORR is dependent on the shape; hence preparation methods and conditions are vital for metal catalysts for ORR [31]. Figure 3 is a schematic illustration of various shapes formed by metal nanostructures depending on the preparation methods, conditions and metal properties. As the essence of a synthesis, a metal precursor is reduced to produce metal atoms, which subsequently aggregate to form nuclei. Once the nuclei have grown past a certain size, they become seeds with a single-crystal, single-twinned, or multiple-twinned structure depending on the prepation methods and conditions used. If stacking faults are involved, the seeds will grow into plate-like nanostructures. The parameter R is defined as the ratio between the growth rates along the <100> and <111> axes [43].

Heat treatment has been recognised as the important or necessary step for ORR activity, as it can induce particle size growth and better alloying degree changes in catalyst surface morphology from amorphous to more ordered states, all of which have significant improvement in ORR activity and stability [44].

Electrocatalysts for ORR were not only focused on the core-shell structures, but various nanostructures of metal catalyst have also been investigated, such as nanocubes, nanotubes and nanodendrites.

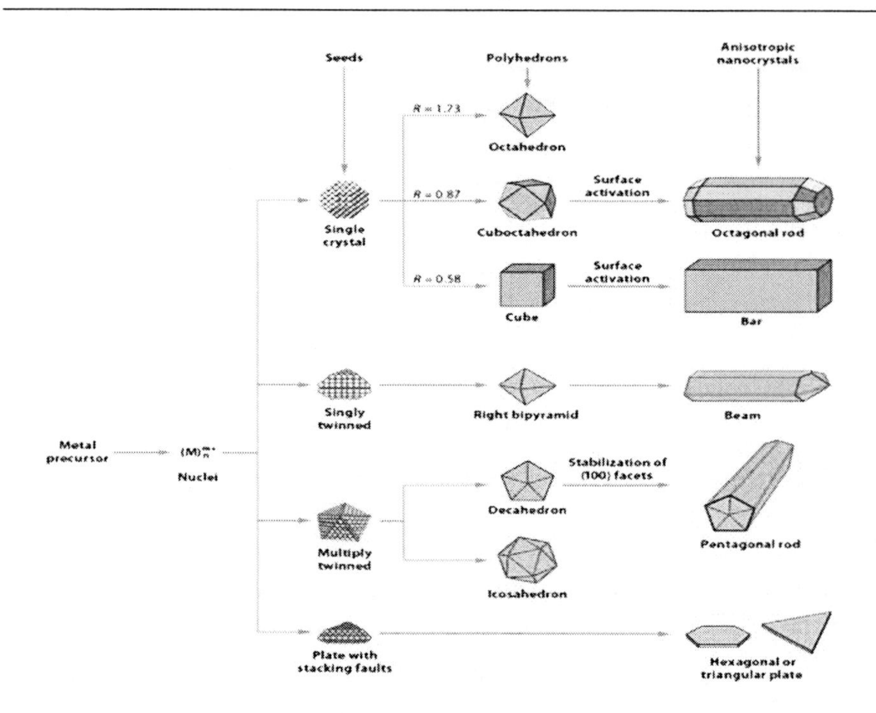

Figure 3: A schematic illustrating various shapes formed by metal nanostructures depending on the preparation methods, conditions and metal properties.[43] ["Reprinted with permission from Wiley-VC"]

Various researchers have reported on the synthesis of various nanostructures of metal catalysts and their electrocatalytic activity compared to the core-shell metal catalyst. Kuai et al. [44] reported uniform, high-yield (>90%) icosahedral Ag and Au nanoparticles by using a hydrothermal system. The Au and Ag icosahedral showed excellent stability and much higher electrocatalytic activity than spherical and Pt/C. They also reported positive shifts in the reduction peak potential for oxygen of 0.14 and 0.05 V, while the reduction peak currents of the Ag and Au icosahedral were 1.5- and 1.6-fold, respectively. Wang et al. [45] reported a simple method to monodisperse Pt nanoparticles with a nanometre size range (3-7 nm) and controlled polyhedral, truncated cubic or cubic shapes; they also studied ORR activities. In their synthesis, the sizes and shapes of the Pt nanoparticles were controlled by reaction temperature. Their results showed that the current density from the ORR for Pt nanocubes is four times that of polyhedral Pt or truncated cubic Pt nanoparticles, showing that ORR activity is dependent on the shape, not on the size, of the Pt nanoparticles. Work on Pt nanocubes was also done by Alia [46], whose results showed higher stability and activity towards ORR compared to Pt/C.

Yu et al. [47] fabricated Pt concave nanocubes enclosed by high-index facets, including (510), (720) and (830). The Pt concave nanocubes showed significantly enhanced ORR activity relative to that of Pt nanocubes, cuboctahedral and commercial Pt/C catalysts that are bounded by low-index facets such as (100) and (111).

Further, Liang et al. [48] used a templating method to prepare free-standing Pt nanowires. The free-standing nanowires showed high stability compared to Pt/C and Pt black. ORR is enhanced through free-standing Pt nanowires because these nanowires possess many novel structural characteristics, including flexibility, large area per unit volume, high stability, preferential exposure of highly active crystal facets and easy electron transport. Although these different forms of Pt nanostructures showed improved activity and stability, the costs and agglomeration of these catalysts contribute to the performance decay of DMFC.

For binary catalysts, ORR improvement depends on the second metal. PtM/C binary catalysts are classified in four categories: highly corrosive and highly active (M = Fe, Co, V and Mn), corrosive and highly active (M = Zn, Cu, Mo and Ni), stable but less active (M = Zr, Cr and Ta) and stable and active (M = W and Ti). Many reports on Pt alloying with transition metals such as V, Co, Cr, Ni, Fe and Cu and these bimetallic catalysts have shown improvement in electrocatalytic activity towards ORR compared to Pt alone in low-temperature fuel cells [5, 36, 45, 49, 50], but the lack of procedures for controlled large-scale synthesis has limited the use in commercial devices [45]. Nishanth et al. [50] and Lim et al. [51] prepared PtPd bimetallic catalysts. They reported that the PtPd electrocatalyst exhibits higher electrocatalytic activity towards ORR in the presence of methanol compared to the commercial Pt/C catalyst. Sarkar et al. [52] synthesised PdW with Vulcan XC-72 carbon. Low-cost PdW is found to increase both electrocatalytic activity for ORR and the catalyst durability.

Furthermore, PdW catalysts offer high tolerance to methanol compared to Pt.

The amount of Pt was further reduced by alloying with the third late transition metal (Au, Pd, Ir, Ru, Rh, Re or Os). Several of these mixed Pt monolayers deposited on a Pd single crystal or on carbon supported Pd nanoparticle exhibited up to 20-fold increases in ORR activity when compared to a Pt-only electrocatalyst. Density functional theory (DFT) calculations show that the higher activity originated from between the Pt monolayer and the Pd substrate and from reduced OH coverage on Pt sites, the result of enhanced destabilisation of Pt-OH induced by the oxygenated third metal.

The activity enhancement observed when using supported Pt alloy can be attributed to interaction between metals which can significantly decrease the Gibbs free energy for the electron transfer steps in the ORR, resulting in enhanced ORR kinetics.

Researchers have also worked on non-Pt catalysts such as Pd, Rh and Au as cathode catalysts for ORR in an attempt to reduce costs. Among these, Pd is preferred due to its selective ORR activity in the presence of methanol. Pd possesses ORR activity very close to that of Pt. It is well documented that alloying of Pd catalysts with transition metals improves their ORR activity, owing to changes in Pd-Pd bond length, modification of the electron configuration and alteration of surface species and compositions.

Wang et al. [53] synthesised PdCo using a modified polyol method followed by heat treatment at temperatures of 400 to 700 °C, while Zhang et al. [2] synthesised PdCo using an impregnation method followed by heat treatment at temperatures of 350 to 700 °C. Both reported a decrease in lattice parameter, an increase in particle size and an increase in the degree of alloying that lead to high ORR electrocatalytic activity [11, 41]. PdCo with a 90: 10 composition also showed the highest ORR electrocatalytic activity, about double of that of carbon-supported Pt, while compositions containing more than 20% Co showed decreased activities [54]. Fernandez et al. [55] associated the improvement of ORR activity to dissociation of O-O bonds and thereby the produced Co-O species, which might transfer to the Pd site in which ORR could take place immediately. Among the alloying of Pd with different transition metals, there has been conflicting results on which binary catalysts show the best ORR activity. It was concluded that the catalyst with best activity depends on the different alloying degree and also on the particle size of various Pd metal In the presence of methanol, PdNi showed the highest ORR activity compared to Pd. The PdCo is reported to show the most positive shifts of the ORR onset potential compared to Pd.

For Au nanoparticles, using DFT has indicated that the oxygen binds to both Au (111) and (211) surfaces. From this study, DFT also indicates that steps and tensile strain facilitate oxygen activation on gold surfaces. Bi also showed an almost two-fold increase in reduction current in an alkaline solution.

AuPd core-shell nanostructures with well-defined morphologies are reported as exhibiting excellent electrocatalytic activity and stability for many Pd catalysed reactions. Fu et al. [41] reported AuPd/CNT with higher ORR activity and stability than commercial Pd black and Pt black in alkaline media, due to unique porous surface structures and synergistic effects between the Au core and the Pd core.

A review by Wang [56] also focused on non-Pt catalysts for ORR. He reports on Ru as the noble catalysts, for ORR tested in an alkaline solution. The results indicated the formation of ruthenium oxide. The reaction processed in a two-way reaction where hydrogen peroxide was formed.

For non-noble catalysts, various metals, transition metals, metal oxides and alloys have been tested for ORR. Wang's review [56] focuses on various non-noble catalysts, such as Cu, Ni, TiO_2, vanadium oxides, WC and $LaMnO_3$ for ORR. None of the non-noble catalysts discussed in his review showed higher catalytic activities compared to Pt. He also concluded that the contributing factor for high electrocatalytic activity was the size distribution of catalysts on the electrode surface.

Other metal oxides such as TiO_2, IrO_2, NiO, CeO_2, ZrO_2 and SnO_2 were investigated for ORR [57], as they are able to provide reversible pathways for ORR. Most of them were found to be unstable in ORR under acidic conditions.

The reduction of oxygen takes place at the cathode and proceeds through the following steps in an acidic solution:

$$O2 + 4H^+ + 4e- \rightarrow H_2O \quad (20a)$$

$$O2 + 2H^+ + 2e- \rightarrow H_2O_2 \quad (20b)$$

$$H_2O_2 + 2H^+ + 2e- \rightarrow 2H_2O \quad (21)$$

The overall reaction is as follows:

The reduction of oxygen takes place at the cathode and proceeds through the following steps in an alkaline solution:

$$3/2O_2 + 3H_2O + 6e- \rightarrow 6OH- \quad (22)$$

$$CO_2 + 2OH- \rightarrow CO_3^{2-} + H_2O \quad (23)$$

The overall reaction is as follows:

$$CH_3OH + 8OH- \rightarrow CO_3^{2-} + 6H_2O + 6e- \quad (24)$$

The mechanism for the reduction reaction for a Pt surface is a multi-electron process with a number of elementary steps, involving different reaction intermediates. Below shows the ORR mechanism of Pt investigated by theoretical calculation based on the electronic structure using DFT [58]. The dissociative mechanism and associative mechanism are proposed for a low current density range and high current density range, respectively:

Dissociative mechanism:

$$1/2O_2 + Pt \rightarrow O\text{-}Pt \quad (25)$$

$$O\text{-}Pt + H^+ + e- \rightarrow OH\text{-}Pt \quad (26)$$

$$OH\text{-}Pt + H^+ + e^- \rightarrow H_2O + Pt \tag{27}$$

In this mechanism, no H_2O_2 is produced. On a Pt surface, O_2 adsorption breaks to the O-O bond and forms adsorbed atomic O, which further gains two electrons in the two consecutive steps, to form water. Because there is no adsorbed O_2 on the Pt surface, H_2O_2 cannot be formed. This mechanism is a detailed four-electron pathway.

Associative mechanism:

$$O_2 + Pt \rightarrow O_2\text{-}Pt \tag{28}$$

$$O_2\text{-}Pt + H^+ + e^- \rightarrow HO_2\text{-}Pt \tag{29}$$

$$HO_2\text{-}Pt + H^+ + e^- \rightarrow H_2O + O\text{-}Pt \tag{30}$$

$$O\text{-}Pt + H^+ + e^- \rightarrow OH\text{-}Pt \tag{31}$$

$$OH\text{-}Pt + H+ + e^- \rightarrow H_2O + Pt \tag{32}$$

This mechanism also does not involve H_2O_2. As adsorbed O_2 is present, O-O may not be broken down in the following steps, resulting in the formation of H_2O_2. H_2O_2 could either be further broken down to H_2O or be the final product.

In a review article by Bezerra et al. [59] the effects of heat treatment for ORR are discussed. He reported on work done by Watanabe et al. and Han et al. where they confirmed that heat treatment had no effect on particle size, but that different electrocatalyst crystallinities could be observed at different temperatures. The same report was published by Valisi et al. [60]. Some researchers still believe that the ORR activity of Pt nanoparticles is dependent on particle size, that the ORR decreases with the Pt particle size. A maximum activity per unit mass of Pt is said to be at the average particle size of 3 nm to 5 nm [59].

Takasu et al. [61] attributed the particle size in ORR to a stronger interaction of oxygen with smaller Pt particles, that is, an electron effect.

Later, Watanabe et al. [62] came up with a different explanation for the observed particle size effect. According to the so-called territory theory, namely that oxygen diffusion to the Pt surface can be influenced by nearby Pt particles, consequently not all of the Pt surface area is utilisable for the reaction. The particle size effect is therefore not truly dependent on the crystallite size, but rather on the inter-particle distances, and the specific activity was reported to be constant at a critical distance of about 18 nm and above.

Higuchi et al. [63] reported that the Pt particle distribution on carbon support would not affect the specific ORR activity. Instead, the ORR can be affected by the diffusion effect induced from a thick catalyst layer. All these results suggest that the ORR activity can be subjected to significant influence from the mass transfer limitation.

Considerable work has been done on trying to resolve whether ORR is particle size-dependent and/or particle distribution-dependent. It was suggested that ORR may be a structure-sensitive reaction. Shih et al. [64] also reported that differently prepared Pt catalysts have similar intrinsic activity, and that the ORR over Pt is a structure-insensitive reaction.

A recent study by Shao et al. [65], however, showed that ORR is dependent on particle size. They reported that mass activity and specific activity are Pt particle size-dependent. The specific activity increases rapidly (four-fold) as particle size grows from 1.3 to 2.2 nm and increases slowly as particle size increases further, while the maximum mass activity was observed at 2.2 nm. They concluded that the particle size-dependent behaviour is associated with oxygen-binding energies on the different Pt sites accessible on cuboctahedral particles of various sizes. DFT calculations performed on nanoparticles show that the (111) facets contribute to the high activity observed on the 2.2 nm Pt particle due to a proper oxygen-binding energy. Higher binding energy at edge sites and even the (111) facet on a smaller particle result in much lower oxygen reduction kinetics.

PREPARATION METHODS

The preparation methods have an influence on the mean particle size, particle size distribution, the bulk and surface of catalysts composition, the oxidation state of catalysts, the extent of catalyst alloying, the distribution of catalyst crystal surfaces, the catalyst morphology, etc. [28, 66], hence on the electrocatalytic activity of the metal catalysts [15]. The benchmarks for high-performance catalysts include: 1) a narrow nanoscale size distribution; 2) a uniform composition throughout the nanoparticles; 3) a fully alloyed degree; and 4) high dispersion on carbon support [67]. According to these standards, innovative and cost-effective important preparation methods have been developed that show promise for reaching performance optimisation by controlling synthetic procedures and conditions. These methods include impregnation, colloidal and micro-emulsion methods, which are the three important methods for preparing supported PtRu catalysts for DMFC. The impregnation and colloidal methods [5] are known as wet chemistry synthesis methods, while micro-emulsion is the latest developed method that uses micro-reactors. All these methods include a chemical step for forming nanoparticles and a deposit step for dispersing the catalyst onto the carbon particles.

The Impregnation Method

Of these three methods, impregnation is the most used preparation method, because the metal composition is pre-controlled, hence advantageous to use especially when preparing stoichiometric and homogenous catalysts with a narrow size distribution. The impregnation method consists of an impregnation step followed by a reduction step. During the impregnation step the catalyst precursors are impregnated with the support material by mixing the two in an aqueous solution to form a homogenous mixture. This is followed by a reduction step, which is required to reduce the catalyst precursor to its metallic state [68].

During reaction, the catalyst support plays a major role in terms of penetrating and wetting the precursors and it can also limit the nanoparticle growth. The reduction step can be chemical or electrochemical and is performed under high temperatures. Common liquid phase-reducing agents include $Na_2S_2O_3$, $NaBH_4$, $Na_2S_4O_5$ and N_2H_4, and a predominant gas phase-reducing agent is hydrogen. During the impregnation process, many factors can affect the dispersion, morphology and composition of catalysts, causing change in electrocatalytic activity, hence synthetic conditions including the attributes of the metal precursor used, the reduction method and heating temperature are important. The porosity of the catalyst support can adequately control the catalyst nanoparticle size and dispersion.

Metal chloride salts (e.g., H_2PtCl_6 and $RuCl_3$) are normally used as precursors during the impregnation process due to their easy availability. However, it has been disputed that the metal chloride salts might lead to chloride poisoning, reducing the dispersion degree, electrocatalytic activity and stability of the catalyst. In order to reduce chloride poisoning, chloride-free metals such as metal sulphite salts ($Na_6Pt(SO_3)$ and $Na_6Ru(SO_3)_4$) and other Cl-free compounds such as $Pt(NH_3)_4(OH)_2$, $Pt(C_8H_{12})(CH_3)_2$ and $Ru_3(CO)_{12}$ have been used. A number of researchers have reported on metal precursors containing nitrate and carbonyl, giving high mass-specific current density with higher dispersion compared to those containing chloride with carbonyl complexes, showing the best performance towards electrocatalytic activity [69- 73].

A modified impregnation process has been developed where a single precursor is used for multi-component catalysts instead of using different metal precursors, e.g., $PtRu_5C(CO)_{18}$ and $Pt_2Ru_4(CO)_{16}$. Guo et al. [74] reported a modified impregnation method to prepare a well-dispersed PtRu/C that gave a power density of 44 $mWcm^{-2}$ in a single DMFC testing at 70 °C, which was comparable to 42 $mWcm^{-2}$ of commercial E-TEK catalyst with the same composition at similar testing conditions. The drawback with this method is that it is too complicated and uses too many organic compounds, which limit its application. Also, in order to reduce the effect of thermal treatment on catalyst size, the microwave dielectric loss heating has also been introduced in the impregnation process.

Another drawback of the impregnation method is the difficulty in controlling nanoparticle size and distribution, which is related to using liquid solutions as the processing medium, although a highly dispersed catalyst can be obtained by carefully controlling appropriate preparation conditions.

The Colloidal Method

In recent years, the colloidal method has become another explored catalyst synthetic route for DMFC. It includes the following common steps: 1) preparation of metal colloids (precursors); 2) deposition of the colloids into the support; and 3) chemical reduction of the mixture. In the colloidal method, the size of the metal nanoparticles is controlled or stabilised by protecting agents, such as ligands, surfactants or polymers. There are several colloidal routes that have been developed with different preparation conditions, reducing agents and stabilisers. The review by Liu et al. [54] discusses different colloidal routes, including the work done by Watanabe et al. who prepared a highly dispersed PtRu catalyst through decomposition of colloidal Pt and Ru oxides on carbon in aqueous media, followed by a reduction with H_2. The metal oxide colloidal route produced catalysts with a higher specific surface area compared to catalysts prepared using the impregnation method.

Liu et al. [54] also mention work done by Bonnemann et al. in preparing catalysts using an organometallic colloidal route by stabilising the metal precursors with organic molecules, resulting in easy control of particle size and distribution. Bonnemann et al.'s route consists of three steps: 1) pre-forming surfactant-stabilised metal colloids; 2) adsorbing the colloids on the support; and 3) removing the organic stabiliser shell by thermal treatment in an O_2 and H_2 treatment, respectively. By this route the PtRu catalysts were well-defined, completely alloyed particles and a very narrow particle size distribution (< 3 nm) was obtained with comparable activity (~70 mAg^{-1}(PtRu) at 500 mV (vs. RHE) in 0.5 M H_2SO_4 + 0.5 M CH_3OH at 60 °C compared to the commercially available catalyst

(~80 mAg^{-1}) at similar testing conditions. Organometallic stabilisation was also integrated with the metal oxide colloidal route to prepare surfactant-stabilised $PtRuO_x$ fine colloids with a narrow size distribution of 1.5 ± 0.4 nm [75]. However, the organometallic colloidal method is not suitable for large-scale production due to the costly surfactants and requires a substantial number of separation and washing steps.

Other colloidal routes using various reducing agents, organic stabilisers or shell-removing approaches such as dodeclydimethyl (3-sulfo-propyl) ammonium hydroxide (SB12) and polyvinylpyrrolidone (PVP) as the stabiliser have been developed [76, 77]. The tuneable colloidal method is simple and very promising in the synthesis of catalysts.

Recently, graphene oxide (GRO) and single-walled carbon nanotubes (SWCNTs) have been used as reducing agents using a pyrolysis process. The modified pyrolysis method has been reported by Li et al. [73], where NaOH, ethylene glycol and carbon black were used as reducing agents. Compared to the commercial Pt/C and Pd/C, the as-prepared Pd/C catalyst shows superior methanol tolerance as the cathode catalyst in DMFC.

Xue et al. [77] reported a two-step pyrolysis process to prepare highly dispersed PtRu catalysts. The Pt and Ru precursors were combined with ethylene glycol and carbon Vulcan XC-72 to form an aerosol by an atomiser. The aerosol droplets were passed through a heated quartz tube in which PtRu nanoparticles formed on carbon support by solvent evaporation and precursor decomposition. The prepared 30% PtRu/C catalyst showed a cell voltage of 0.39 V compared to 0.34 V obtained from the commercial catalyst (E-TEK) at a current density of 300 mAcm-2 in a DMFC testing at 90°C.

Ethylene glycol has also been used as both solvent and reducing agent. This method is known as the modified polyol method. In this process, a polyol (e.g., ethylene glycol solution) containing the metal precursor salts is refluxed at about 120 °C, where polyol decomposes homogenously, releasing the reducing agent for ion metal reduction. Bensebaa et al. [78] prepared a PtRu catalyst using ethylene glycol as both the solvent and the reducing agent and PVP as the stabiliser. Lim et al. [51] fabricated Pt/C using a modified polyol method, where the metal precursor was mixed with a treated ethylene glycol-carbon mixture. High activity of Pt/C for ORR was obtained and the mass activity was 1.7 times higher when using modified polyol method compared to commercial Pt/C.

Recently, there have been reports where microwave heating has been used to prepare high-purity metal nanoparticles using what is known as a microwave-assisted polyol process. Microwave irradiation through dielectric heating loss is fast, simple, uniform and energy-efficient [79]. Liu et al. [72] developed a glycol polyol colloidal route in which an organic stabiliser was unnecessary. PtRu colloids were prepared in an ethylene glycol solution and transferred to a toluene medium with decanethiol as a phase transfer agent. The microwave was used to remove the organic shell from those colloids. The fast heating by microwave accelerates the reduction of the metal precursor and the nucleation of metal clusters.

The advantages of using the colloidal method compared to the impregnation method are: a) The size of the metal particles can be tailored independently of the support; b) due to the stabilising or protecting, shell particle sizes smaller than 2 nm can be obtained as well as uniform particle size distribution; and c) a good catalyst dispersion even for high noble metals loadings and appropriately high surface areas can be obtained. The disadvantages are: 1) The stabilising/protecting shell blocks some part of the metallic surface and appropriately decreases the available surface area for the electrochemical reaction; 2) a protecting agent is present, which can be difficult to remove once the particles are adsorbed onto the support and can obstruct the catalytic performance of the nanoparticles; and 3) the overall synthesis involves a higher price and complexity.

The Micro-Emulsion Method

Micro-emulsion is the latest method to be developed for synthesis of PtRu. It involves the formation of metal nanoparticles through a water-in-oil micro-emulsion reaction, followed by the reduction step. The micro-emulsion serves as the nanoscale reactor where the chemical reaction takes place with nanoscaled aqueous liquid droplets containing a noble metal precursor. The droplets are absorbed by the surfactant molecules and uniformly dispersed in an immiscibly continuous organic phase. The reduction step is processed either by adding a reducing agent (e.g., N_2H_4, HCHO or $NaBH_4$) into the micro-emulsion system, or by mixing it with another reducing agent containing the micro-emulsion system. As a consequence, the reduction reaction is enclosed to the inside of the micro-emulsion reactor, and the produced metal particle sizes can be easily controlled by the significance of the micro-emulsion size. The surfactant molecules can behave as protective agents to prevent the nanoparticles from agglomeration. The surfactant molecules can be easily removed by heat-treating the supported nanoparticles [80-84].

The PtRu prepared by this method were found to produce higher electrocatalytic activity than the commercially available catalysts for methanol oxidation in both an electrochemical half-cell reaction containing sulphuric acid and methanol oxidation in DMFC. The main advantage of the micro-emulsion method is its simplicity in controlling metallic composition and particle size within a narrow distribution by changing the synthetic conditions.

Zhang and Chan [85] synthesised PtRu/C catalysts by a two micro-emulsion route, in which the metal precursor and reducing agent formed two separate micro-emulsion systems. Their catalysts produced a very narrow size distribution (2.5 ± 0.2 nm) and were highly alloyed. Xiong and Manthiram [84] reported that the nanoparticle size depends on the ratio of water to surfactant, namely sodium bis-(2-ethylhexyl)-solfosuccinate (AOT) in their instance. They found that the PtRu/C nanoparticle size increased initially with increasing water and became almost constant at water >10. These results showed that a mono-size distribution could be easily obtained by controlling the synthetic conditions.

However, as with the organometallic colloidal method, the micro-emulsion is costly due to its surfactant molecules and a series of separation and washing steps, hence not suitable for large-scale production [83].

SUPPORTS FOR DMFC

Catalysts supports not only play a role in stabilizing electrocatalysts, but are also very critical in mass transfer management in fuel cells and providing a physical surface for accomplishing a high surface area [1]. The performance and stability of DMFCs depend on the interaction of metal particles and support material influencing the dispersion of electrocatalysts. Surface area, porosity, electrical conductivity, electrochemical stability and surface functional groups designate a support [84, 85]. An absolute support should offer the following: 1) good electrical conductivity; 2) good catalyst-support interaction: 3) large surface area; 4) mesoporous structure enabling the ionomer and polymer electrolyte to bring catalyst nanoparticles close to the reactants; 5) good water-handling capability to avoid flooding; 6) good corrosion resistance; and 7) easy recovery of catalyst.

The carbon particles are usually used as catalyst supports due to having several advantages over unsupported catalysts, including 1) higher stability than unsupported catalysts in terms of agglomeration under fuel cell operating conditions in both acidic and basic media; 2) the porosity of the carbon black support allowing for gas diffusion to the active sites; 3) the good electronic conductivity of the carbon support providing electron transfer from catalytic sites to the conductive carbon electrodes and then to the external circuit; and 4) the small dimensions of nanoscale catalyst particles dispersed on a carbon support maximising the contact area between catalyst and reagents [5, 83, 84, 86].

Carbon Black

For decades, highly conductive carbon black has been used as supports for DMFCs. There are different types of carbon black, including Vulcan XC-72, acetylene black, ketjen black, etc. and these are mostly synthesised by hydrocarbon pyrolysis from natural gas or oil fractions taken from petroleum processing. These carbon blacks show different physical and chemical properties, such as specific surface area, porosity, electrical conductivity and surface functionalisation. Among these factors, specific surface area has the significant effect on the preparation and performance of the supported catalysts. Acetylene black carbon support is the one with the least specific surface area, hence it is not possible to prepare highly dispersed supported catalysts. High surface area kejten black is able to support highly dispersed catalyst nanoparticles. However, highly dispersed ketjen black catalysts show high ohmic resistance and mass transportation limitation during fuel cell operation. High surface area ~250 m^2g^{-1} with high mesoporous distribution and graphite character, in this case Vulcan XC-72, has been mostly used as carbon support for the preparation of fuel cell catalysts because of its good compromise between electronic conductivity and BET surface area [82]. Issues with carbon black include side reactions that give H_2O_2, low surface area compared to nanostructured materials and corrosion [87]. Much higher surface area carbon supports, including nanostructured and mesoporous carbons, which are discussed below, have been proposed as catalyst supports for DMFC.

Nanostructured Carbon

Nanostructured carbon supports have been investigated as possible best supports for DMFC catalysts due to their superior properties, namely 1) specific surface area; 2) porosity; 3) morphology; 4) surface functional group; 5) high electron conductivity; and 6) corrosion resistance [88, 89].

Carbon nanotubes (CNTs) are the best-known nanostructured carbons, which have shown very promising results in catalyst support for DMFCs due to their unique electrical and structural properties. In addition, they have a very high crystallite status, which makes them highly conductive [4, 90, 91]. The reported studies have shown that CNTs are superior as catalyst supports compared to carbon black for DMFC.

CNTs are 2D nanostructures, tubes formed by rolled-up single sheets of hexagonally arranged carbon atoms. They can be single-walled carbon nanotubes (SWCNTs) or multi-walled CNTs (MWCNTs), depending on the structure. A SWCNT is a graphene sheet rolled into a cylindrical shape and an MWCNT consists of coaxially arranged graphene sheets rolled into a cylinder. Both SWCNTs and MWCNTs have been extensively studied for DMFC. MWCNTs have been found to be more conductive, while SWCNTs provide larger surface areas.

CNTs are promising preferred supports for DMFCs. However, the CNT synthesis, metal loading and electrode preparation still present challenges, especially for fuel cells preparation. CNTs can be synthesised using carbon-arc discharge, laser ablation of carbon and chemical vapour deposition. These synthesis methods have their limitations in terms of large-scale synthesis and cost-effectiveness. It is necessary to further develop industrial large-scale production of CNTs to meet the needs of all applications, including fuel cells.

Deposition, distribution and crystallite size of metal nanoparticles supported on CNT are significantly affected by factors such as synthesis method, oxidation treatment of CNT and the metal precursors [92]. Several methods have been developed to prepare highly dispersed metal/CNT catalysts. It was also found that the surface modification of CNTs before metal deposition was important for obtaining optimal interaction between the support and the metal precursor. CNTs are said to be chemically inert, hence it is difficult to attach the metal nanoparticles to the substrate surface [93]. Through surface modification or pre-treatment of CNTs, some binding sites are introduced onto the surface so that the metal nanoparticles can easily attach onto the CNT surface [54, 94]. The effects of pre-treatment include the introduction of functional groups of support such as -OH, -COOH, -C = O, etc., change in electrocatalyst dispersion and morphology, and crystallite growth on the carbon support. The most commonly used pre-treatment includes refluxing CNTs in strong acids such as nitric acid and sulphuric acid to create an acid site on the surface, which can act as a nucleation centre for metal ions. CNTs can also be modified using other forms such as sulphur and conjugated polymers. With pre-treatment, better dispersion of metal catalyst nanocatalyst, better size control and distribution have been reported.

Surface-modified CNTs have been used to support a wide variety of mono-, binary as well as ternary catalysts. Another challenge of CNTs as catalyst support for DMFCs is how to use them to fabricate high-performance working electrodes. The process of microstructure formation depends on the type of the supported catalyst, the type of ionomer added and the type of dispersion medium used during ink preparation. If the catalyst was prepared by the usual ink process, it was estimated that only 20 to 30% Pt catalyst was utilised because of the difficulty of reactants to access inner electrocatalytic sites. Developing new electrode structures could make use of CNTs' advantages of electrical, mechanical and structural properties.

Considerable work has been done on CNTs [95-97] supported on catalysts, showing better performance compared to those supported on carbon black Vulcan XC-72, including higher electrocatalytic activity towards methanol oxidation and oxygen reduction and 20 times higher current density under the same conditions in both half-cell characterisation and fuel cell performance tests. The higher electrocatalytic activity was attributed to several factors: 1) the prepared catalysts, i.e., PtRu/MWCNT, possess better nanoparticle sizes and ideal compositions that improve their electrocatalytic activity; 2) the functional groups formed on the CNT surfaces result in high hydrophilicity, which creates better electrochemical reaction environments on the electrode; and 3) the high electronic conductivity of MWCNT lowers the resistance in methanol oxidation to a larger surface area and lower overpotential of CNTs for methanol oxidation reaction.

Park et al. [96] performed accelerated stress tests on Pt supported on MWCNT to study the polarisation losses and electrochemical behaviour for a PEM fuel cell cathode. When compared to Pt supported on Vulcan XC-72, the MWCNT-based catalyst showed higher retention of electrochemical area, smaller increment in interfacial charge, transfer resistance and a slower degradation of fuel cell performance. This conclusively confirmed the higher corrosion resistance of MWCNT and also a stronger interaction with Pt nanoparticles. It was also observed that the high corrosive-resistant MWCNTs prevented the cathode catalyst layer from severe water flooding by maintaining the electrode surface and hydrophobicity for a long period under continuous anodic potential stress.

The cyclic voltammetry of methanol electro-oxidation on Pd/C, Pd/MWCNT, PdNi/C and PdNi/MWCNT electrodes is displayed in Figure 4. The Pd/MWCNT and PdNi/MWCNT cyclic voltammograms show that the potential beginning of the methanol electro-oxidation (onset potential) was -0.45 V vs. Ag/AgCl; while that of the Vulcan XC-72-based catalysts was -0.25 V vs. Ag/AgCl. In addition, a higher current density towards methanol electro-oxidation for MWCNT-based catalysts in alkaline medium was observed, confirming that CNT-supported electrocatalysts are more efficient than C-supported electrocatalysts. For chronoamperometry, it can be seen that the current density for methanol oxidation on the PdNi/MWCNT catalyst is larger than that of Pd/MWCNT, Pd/C and PdNi/C, respectively, in the whole time range. This indicates that the electrocatalytic activity and stability towards methanol electro-oxidation of the catalysts supported on MWCNTs is better than the Pd supported on Vulcan XC-72.

Tang et al. [97] developed some techniques to grow CNTs on fuel cell carbon paper fibres to produce a 3D nanotube-based structure. Pt and Pt alloys are expected to deposit directly onto these novel CNT-based catalyst supports, acting as both gas diffusion layer (GDL) and catalyst layer, which seems to be a promising way to make low-cost electrodes for DMFCs with high maximum power density of 595 $mWcm^{-2}$ using the catalyst loading of 0.04 $mgcm^{-2}$ compared to Pt/Vulcan XC-72 based electrode with 435 $mWcm^{-2}$ with equal Pt loadings and the reference electrode with Pt on hybrid CNT/carbon black blend layer producing 530 $mWcm\text{-}^{2}$.

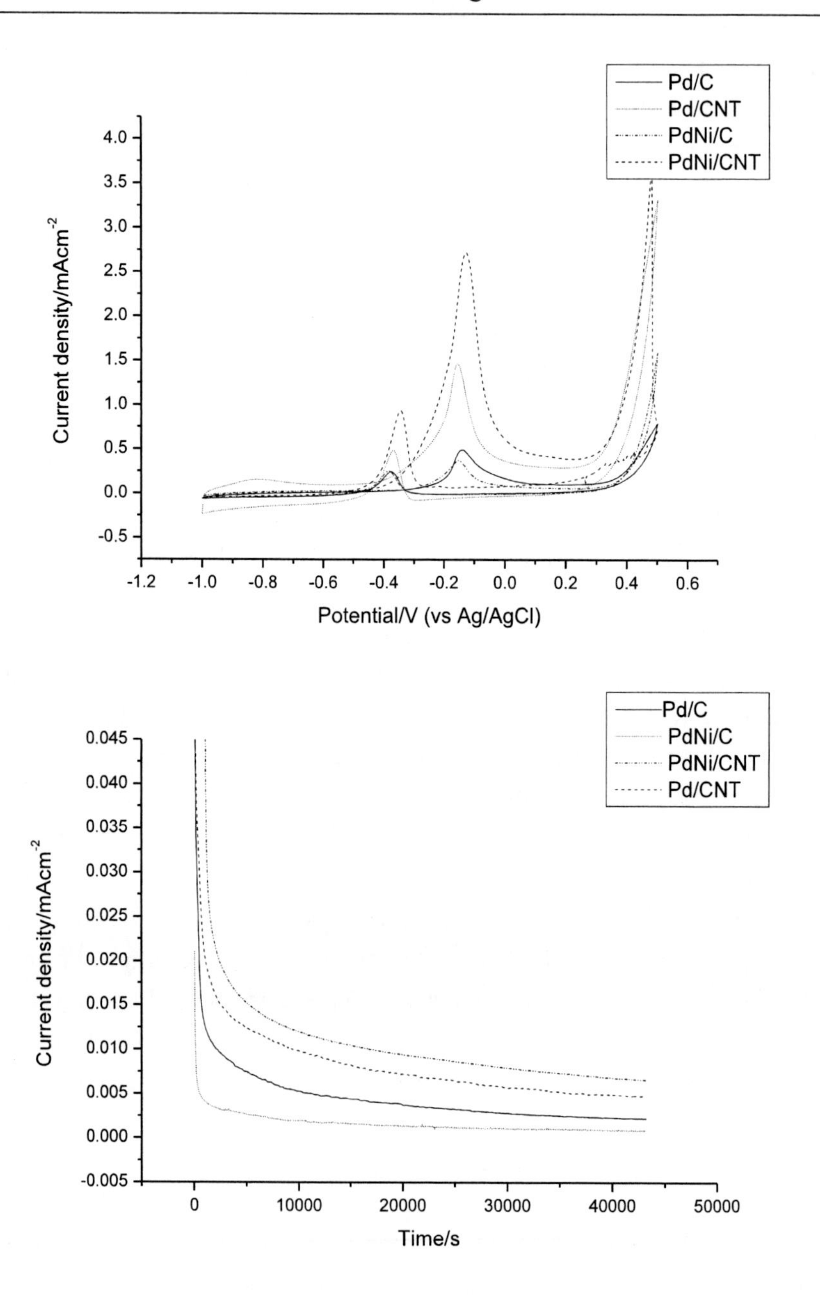

Figure 4. The cyclic voltammetry and chronoamperometry of methanol electro-oxidation on Pd/C, Pd/MWCNT, PdNi/C and PdNi/MWCNT catalysts with a solution of 1 M methanol and 1 M KOH at 25◦C. Scan rate: 30 mVs−1.

Modifying the carbon materials by different N-functional groups is known to enhance the electrocatalytic activity towards ORR in acidic and alkaline media [14]. Fang [98] and Nagaiah et al. [99] reported on Pt nanoparticles on N-doped CNTs synthesised at various temperatures from 200 °C to 800 °C. The Pt/N-doped CNTs at 800 °C exhibited the best electrocatalytic activity, greatly enhanced ORR activity, favourable positive onset potential for the ORR and increased reduction currents and high stability in alkaline media. This improvement has been associated with the so-called N incorporation, which could be used to produce preferential sites of CNTs with low interfacial energy for immobilising Pt nanoparticles. Furthermore, nitrogen is also known to efficiently create defects on carbon materials, which increase the edge plane exposure. N-functional groups transform to more thermally stable structures during heat treatment [100].

Other nanostructured carbons, such as carbon nanofibres (CNF), fullerenes and nanocoils, have been used as catalysts supports for DMFC. CNFs have been reported as superb support materials because of the excellent compromise between textural and structural properties, e.g., nanosized diameter, angle of a graphite plane to the fibre axis and ratio of edge to basal atom. Separation of the catalyst from the reaction mixture has also been reported to be less complicated for CNFs compared to the other carbons. CNFs have three graphitic structures (platelet, ribbon/tubular and herringbone). The platelet CNFs involve lower-cost production compared to SWCNTs (approximately 100 times) and are suitable for mass production [101-103].

Bessel et al. [104] used Pt/CNF and Pt/C (Vulcan XC-72) as catalysts for methanol oxidation. They found that catalysts containing 5% Pt supported on platenet and ribbon CNFs produced comparable activities to that catalysed by 25% Pt supported on Vulcan carbon, while herringbone GNF-supported Pt catalysts showed poor electrochemical activity.

Singh and Dempsey [87] reported that Pt/CNF showed better electrocatalytic activity (276 $mAmg^{-1}_{Pt}$) and improved methanol tolerance of Pt/CNF to CO poisoning during MOR compared to Johnson Matthew commercial Pt/C (224.6 $mAmg^{-1}$).

Tsuji et al. [100] deposited PtRu supported on three types of CNFs using a microwave polyol method. The dependence of particle sizes and electrochemical properties on the structures of CNFs was tested. The methanol oxidation of PtRu/CNF measured at 60 °C was 1.7 to 3.0 times higher than that of a standard catalyst on carbon black (Vulcan XC-72) support. Based on these results, the higher electrocatalytic activity was obtained from the platelet CNF, which is distinctive by its edge surface and high graphitisation degree.

The major difference between CNTs and CNFs is that CNFs have a very thin and hollow cavity. The diameters of CNFs are much larger than that of CNTs and may go up to 500 nm, while the length can be up to a few millimetres. In addition, for CNTs a predominant basal plane is exposed, while for CNFs only the edge planes that present potential anchoring sites (platelets and herringbone) are exposed. Some researchers reported low electrocatalytic activity for PtRu/CNF compared to PtRu/C and this has been linked with the preparation method used [86].

Graphene (GR) has generated interest as potential catalyst support for methanol oxidation reactions due to its high electron transfer rate, large surface area, excellent stability and high conductivity [105]. The 2D planar structure of the carbon sheet allows both the edge planes and the basal planes to interact with the catalyst nanoparticles. The rippled but planar sheet structure also provides a very high surface area for the attaching catalyst nanoparticles.

Pt nanoparticles deposited on a graphene nanosheet gave rise to exceptional modification to the properties to the Pt electrocatalysts with very high activity for methanol oxidation, reflecting the advantage of graphene as catalyst support. Reviews by Choi et al. [58], Yoo et al. [106] and Wang et al. [107] reported on the current density of Pt/graphene (0.12 mA/cm^2) being higher compared to that of Pt/C (0.03 mA/cm^2), while the CO adsorption rate of Pt/GR is 40 times less than that of Pt/C. They also reported on better performance of PtRu-reduced graphene oxide (R-GRO) compared to commercial Pt/C, PtRu/GR and Pt/GR, which also showed higher electrocatalytic activity towards methanol oxidation compared to commercial PtRu/C ETEK. They also reported on bimetallic and ternary graphene-supported catalysts (Pt_3Co, Pt_3Cr, PtPd, PtAu and PtPdAu), which showed improved electrocatalytic activity compared to Pt/C. Nitrogen-doped graphene (N-GR) has been shown to yield promising results, especially for the sluggish ORR. The disorders and defects introduced in the graphene stack due to the incorporation of nitrogen are known to act as anchoring sites for the catalyst nanoparticles. Jafri et al. [108] used nitrogen-doped nanoplatelets as Pt supports for electrocatalytic studies. Membrane electrode assembly fabricated using Pt/N-GR and Pt/GR as the ORR catalyst showed a maximum power density of 440 $mWcm^{-2}$ and 390 $mWcm^{-2}$, respectively. The improved Pt/N-GR was attributed to the formation of pentagons due to the incorporation of N in the C-backbone, leading to an increase in conductivity of neighbouring C atoms.

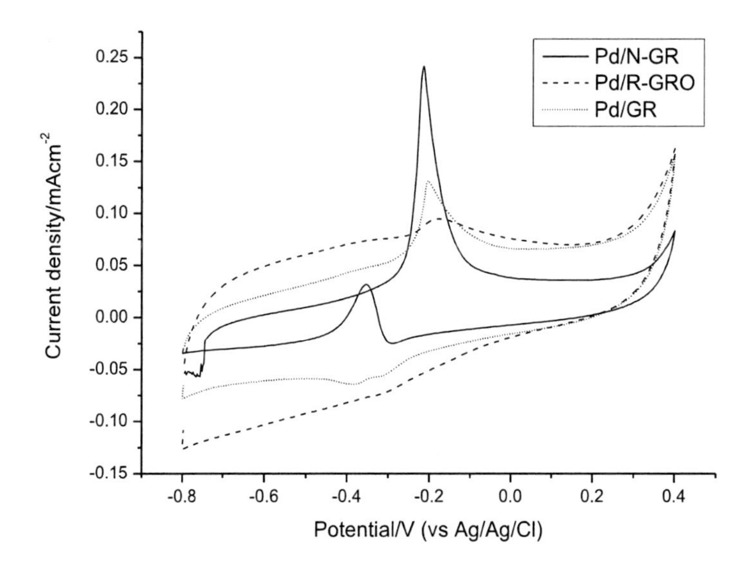

Figure 5. The cyclic voltammetry curves of methanol electro-oxidation on Pd/graphene, Pd/N -graphene and Pd/R-graphene oxide catalysts with a solution of 1 M methanol and 1 M KOH at 25 °C. Scan rate $30mVs^{-1}$.

GRO has also drawn much interest. Although GRO has lower conductivity (a difference of two or three orders of magnitude compared to graphene), it offers a different set of properties, including hydrophilicity, high mechanical strength and chemical ‘tunability,’ compared to graphene [109]. The use of GRO as catalyst support for DMFC is one of the latest applications that have shown promising results. Oxygen groups introduced into the graphene structure during the preparation of GRO create defect sites on surface as well as edge planes. These defect sites act as nucleation centres and anchoring sites for the growth of metal nanoparticles.

Sharma and Pollet [86] prepared (R-GRO) on Pt. The results showed improved electrocatalytic activity. Three mechanisms were attributed to enhancing CO tolerance and methanol oxidation. Firstly, R-GRO promotes water activation due to its hydrophilic nature. Hence, the adsorbed OH species at the Pt edge promotes CO oxidation. Secondly, a strong interaction between Pt and R-GRO was found, which can induce some modulation in the electronic structure of Pt clusters, modifying the Pt-CO binding energy and, as a result, minimizing CO adsorption on Pt. Thirdly, Pt/R-GRO catalyst can promote hydrogen spill-over (the process involving dissociative chemisorption of molecular hydrogen on a supported Pt catalyst surface, followed by the diffusion of atomic hydrogen onto the surface of R-GRO). The R-GRO species on the Pt surface promotes the formation of a -GO species adjacent to CO-poisoned Pt sites, which combine with adsorbed CO to strip it from the surface as CO_2.

Ha et al. [110] reported the use of Pt embedded on R-GRO for ORR in DMFC. ORR studies and single-cell polarisation studies were performed, which demonstrated an 11% higher maximum power density for 70% Pt/R-GRO compared to commercial 75% Pt/C. After long-term stability, the average particle sizes of Pt/C and Pt/R-GRO increased from 4.1 to 5.4 nm and 2.9 to 3.7 nm, respectively, suggesting lower Pt agglomeration, which would encourage mass transport by facilitating reactants to the catalytically active sites.

Other noble metal nanoparticles and bimetallic systems have also been grown on GRO and have exhibited superior electrocatalytic activity. Figure 5 below shows cyclic voltammohrams of GR, R-GRO and N-doped graphene on Pd catalyst were compared for methanol oxidation in DMFC.

The results showed that Pd on N-graphene had higher current density compared to GR and R-GRO.

Nanostructured carbon supports have been alloyed with metal oxides in order to produce highly dispersed Pt nanoparticles by introduction of nucleation sites by metal oxides. Shao et al. [1] reported on Pt/indium tin oxide (ITO)-graphene compared to Pt/graphene towards oxygen reduction. Their results showed that Pt/ITO-graphene performed better than Pt/graphene, while the stability was doubled. This improvement was attributed to the following: 1) ITO nanoparticles protect graphene from corrosion; 2) DFT calculations suggest that, in an oxidising environment, the interaction energy at Pt/ITO-graphene is much higher; and 3) strong metal–support interaction between Pt and SnO_2 (in ITO).

Although nanostructured carbon supports have shown much improvement in terms of corrosion compared to carbon black, the introduction of functional groups using acids compromises durability [111]. Functionalisation using polyelectrolytes (e.g., polydiallydimethyl ammonium chloride and poly(allymine hydrochloride)) have shown improved Pt nanoparticle dispersion and catalytic performance (activity and durability). The positively polyelectrolyte molecules enclose around hydrophobic graphitic carbon due to the dipolar property of these molecules. This improves the dispersion of carbon nanomaterials and provides nucleation sites for Pt nanoparticles due to electrostatic interaction between positively charged molecules and a negatively charged Pt precursor, which is reduced by ethylene glycol.

Mesoporous Carbon

Mesoporous carbon belongs to the class of porous carbon materials with pore sizes of 2 to 50 nm dispensing high surface area and conductivity. They are classified into two categories according to their final structure and method of preparation: 1) ordered mesoporous carbons (OMCs); and 2) disordered mesoporous carbons (DOMCs). OMCs have been widely studied as catalyst supports for fuel cells because they have controllable pore size, high surface areas and large pore volumes [112].

Mesoporous carbons are also known to have some surface oxygen groups that are considered to improve the interaction between the metal catalyst and the support, allowing better dispersion. The mesoporous carbons are known to show better performance compared to carbon black. Salgado et al. [113] synthesised PtRu supported on CMK-3, the most tested mesoporous carbon fuel cell support, to study its performance in DMFC. The performance of the as-prepared catalysts compared with the Pt catalyst on Vulcan XC-72 and commercial PtRu/C from E-TEK was found to demonstrate higher MOR activity compared to Vulcan carbon support. It also showed a shift (~0.15 V), in the CO stripping potential towards more negative values, indicating faster electron transfer kinetics. Ding et al. [114] tested CMK-3 for ORR, and reported that Pt/CMK-3 showed better activity compared to commercial Pt/C.

Similar to other carbon materials, mesoporous carbon has also been doped with nitrogen. Liu et al. [115] prepared N-doped OMC graphitic arrays using a metal-free nanocasting method. A very narrow pore size distribution of approximately 3.8 nm was observed. N-OMC with moderate nitrogen content resulted in a high surface area and a graphitic framework, leading to high electrocatalytic activity, excellent long-term stability and resistance to methanol crossover effects for the ORR compared to commercial Pt/C.

The high surface area of ordered mesoporous carbon (OMC) supports allows high dispersion for DMFC. As compared to carbon black, the OMC provides higher electrocatalytic activity and higher thermal and electrocatalytic stability. The disadvantage of OMC is that it contains small amounts of oxygen surface groups.

Supports such as CNF do not require any chemical treatment due to the presence of highly active edge planes on which the catalyst nanoparticles can be attached.

Non-Carbon Supports

The interest in non-carbon support is mostly due to carbon corrosion ($C + H_2O \rightarrow CO_2 + 4H^+ + 4e^-$), which occurs at potentials higher than 0.207

V (vs. RHE) suffered by carbon supports in different degrees. Corrosion of support leads to issues such as loss of the catalyst, which affects the overall performance of the fuel cell [116]. It is also a severe problem that compromises fuel cell durability. Non-carbon supports (e.g., conductive ceramics such as titanium, tungsten and zirconium oxides) with dimensions in nanometre range have been considered as an alternative due to the fact that they provide the electronic conductivity and the electrochemical stability lost by carbon supports due to corrosion.

TiO_2 has been considered due to chemical and mechanical stability, high corrosion resistance and electrochemical stability in DMFC cathodic potentials in an acidic medium. It has been reported that TiO_2 prevents the agglomeration of Pt particles and effectively disperses Pt atoms in clusters [116]. In addition, TiO_2 supports control the nanostructure of the catalyst and provide both thermal and oxidative stability. TiO_2 is said to show similar performance to carbon-supported Pt, except that it yields lower performance in the ohmic regime due to its higher electrical resistance. Other studies have shown that TiO_2 improves Pt ORR by facilitating mechanisms such as reactant surface diffusion and oxygen spill-over.

Metal oxide composite supports have also been developed to stabilise nanoparticles. However, the oxides suffer from low electrical conductivity. Hence, carbon materials, such as CNTs and graphene, have been introduced into the oxides to improve electrical conductivity and other properties [117]. It is reported that those supports showed increased activity for ORR [118]. Montero-Ocampo et al. [76] prepared Pt/TiO_2-CNT and Pt/TiO_2 using a polyol colloidal method and compared their electrocatalytic activity towards ORR. The data showed that Pt/TiO_2-CNT was more catalytically active than the Pt/TiO_2 due to the limited microscopic electronic conductivity of TiO_2. Wang et al. [56] also studied the performance of Pd nanoparticles on titania nanotubes for MOR. The results showed relative activities were Pd/TiO_2-CNT > PdTiO_2 > pure Pd. Huerta et al. [117] reported on PtTiO_2/C and Pt/C for ORR in acidic media. The ORR activity for carbon oxide-supported catalysts was higher by a factor of four compared to the as-prepared Pt/C. Their finding was attributed to the improved electronic structure and the strong metal substrate interaction, which led to improved ORR electrocatalytic activity.

CONCLUSION

With regard to new fuel cell catalysts for anodes, there are two major concerns: performance, including activity, reliability and durability, and cost reduction.

In recent investigations, Pt nanoparticles are still the promising catalysts for DMFC compared to Au and Pd in acidic media and polymer exchange membranes. PtRu/C is still the most-used catalyst owing to its activity to the polarisation characteristics to MOR. Although alloying Pt with the second or third catalyst enhances electrocatalytic activity has been proven experimentally that the MOR activity and stability could be enhanced through multi-component catalysts [119], the Pt content used is still more than 50%. In addition, the lack of procedures for controlled large-scale synthesis has also limited the use of multi-component catalysts in commercial devices [120].

To this end, PtRu is considered as the best CO tolerance catalyst; however, both metal catalysts are expensive and scarce, hence there is still a need to develop less expensive catalysts with the electrocatalytic activity in MOR comparable to PtRu with higher CO tolerance [121, 122]. It has also been reported that PtRu-based catalysts showed much improved electrocatalytic activity, but the higher activity is less than 10 times as compared to Pt, which is necessary for commercialisation of DMFC. Recently, more research has been focused on Pd metal catalysts, which are five times cheaper than Pt, using the anion exchange membrane and alkaline medium. The results have shown that Pd exhibits better results compared to Pt or PtRu in alkaline electrolytes. The application of alkaline electrolytes to the DMFC could result in a reduction of the catalyst loading and allow the use of non-precious metals catalysts. Hence, Pd and Pd-based catalysts are emerging as alternatives to Pt-based catalysts for alcohol oxidation reactions in alkaline media, although more work is still required.

ORR Pt metal particles are still the most commonly used electrocatalysts due to, among other reasons, the structural complexes of Pt alloys, which have made it difficult to interpret their mechanism.

More recent work has shown that the ORR activity of PtPd is much higher and the dependence of ORR activity in Pd content goes through the maximum, hence platinum-free, Pd or Pd alloys have been proposed as ORR catalysts in alkaline media. Heat treatment with respect to Pt-based catalysts is able to induce particle size growth, better alloying degree and changes in the catalyst surface morphology from amorphous to more ordered states, all of which have a remarkable effect on ORR activity and stability.

In conclusion, noble metals are still the best catalysts for ORR and MOR; however, the high cost and low abundance of noble metal-based catalysts have largely limited their practical applications.

Chemical and physical characteristics of catalysts depend on the preparation procedure, as it is an important factor regarding their electrochemical activity. The two synthetic methods have been improved by using microwave heating. With all the synthesis methods discussed, the major drawback is the costs that reduce the possibility of them to be successfully used in large-scale synthesis.

Carbon support plays an important role in the preparation and performance of catalysts, as it enhances catalyst dispersion and provides an underlying framework for electron conduction and gas diffusion. The performance of the catalyst strongly depends on the nature of carbon support. It has been reported that supports with higher crystallites provide smaller Pt crystallite sizes. The effect is attributed to metal support interaction, which is related to the kind of support and has been reported to affect the growth, structure and dispersion of metal particles.

With the introduction of advanced 1D and 2D carbon supports, such as CNTs and graphene, the problems associated with nanoporosity and low exertion of catalysts have been resolved to a certain extent. The high crystallinity and surface area of CNTs and graphene provide a suitable support for the dispersion of catalyst nanoparticles. Under long-term durability tests, carbon black underwent a morphological transition to induce catalyst aggregation, while no such effects were observed for CNTs. The use of graphene support provides a high electron transfer rate along with high conductivity and the 2D nature allows both the edge and the basal planes to interact with the catalyst. Graphene and the CNT-based supports provide adequate advantages concerning mass and charge transport by providing shorter effective lengths for electronic and ionic support

The use of nanostructured carbon support has improved the performance of the catalyst supports used in DMFC, as they display a strong influence on catalyst durability and behaviour. However, carbon corrosion still exists for these catalysts; although using carbon supports such as CNT has reduced corrosion, it is not completely eliminated. In addition, carbon supports such as CNTs are inert, hence functionalisation of the carbon support is used to improve the anchorage of the catalyst nanoparticles on the support and reduce agglomeration. However, the functionalisation of the carbon support can make it more responsive to electrochemical oxidation, leading to loss of active surface area. It can also affect the ionomer distribution, hence affecting proton conductivity of the fuel cell electrodes.

For nanostructured supports such as CNT, graphene or mesoporous carbon to completely replace carbon black as catalyst supports, further and more extensive testing of fuel cells, especially in terms of electrochemical activity and long-term stability, needs to be accomplished.

REFERENCES

[1] Shao, Y., Cgeng, Y., Duan, W., Wang, W., Lin, Y., Wang, Y., Liu, J. *ACS Cat.* 2015, *5* (12), 7288-7298.

[2] Zhang, L., Lee, K., Zhang, J. *Electrochim. Acta.* 2007, *52*, 3088-3094.

[3] Lima, A., Coutanceau, C., Leger, J. M., Lamy, C. *J. Appl. Electrochem.* 2001, *31*, 379-386.

[4] Wang, Z.-B., Yin, G.-P., Shao, Y.-Y., Yang, B.-Q., Shi, P.-F., Feng, P.-X. *J Power Sources.* 2007, *165*, 9-15.

[5] Liu, H., Song, C., Zhang, L., Zhang, J., Wang, H., Wilkinson, D. P. *J Power Sources.* 2006, *155*, 95-110.

[6] Larsen, R., Ha, S., Zakzeski, J., Marcel, R. I. *J Power Sources.* 2006, *157*, 78-84.

[7] Bagotzky, V., Vassiliev, Y., Khazova, O. *J Electroanal Chem.* 1977, *81*, 229-238.

[8] Mahapatra, S. S., Datta, J. *Int J Electrochem,* 2011, *2011*, 1-16.

[9] Ramya, K., Dhathathreyan, K. S. *J Electroanal Chem,* 2003, *542*, 109-115.

[10] Vidakovic, T. Kinetics of methanol electrooxidation on PtRu catalysts in membrane electrode assembly, dissertation, University of Magdeburg, Magdeburg, 2005.

[11] Antolini, E. *Chempluschem.* 2014, *79*, 765-775.

[12] Gurau, B., Viswanathan, R., Liu, R., Lafrenz, T. J., Ley, K. L., Smotkin, E. S. *J. Phys Chem B.* 1998, *102*, 9997-10003.

[13] Watt-Smith, M. J., Friedrich, J. M., Rigby, S. P., Ralph, T. R., Walsh, F. C. *J Phys D Appl Phys.* 2008, *41*, 174004-174011.

[14] Tiwari, J. N., Tiwari, R. N., Singh, G., Kim, K. S. *Nano Energ,* 2013, *2*, 553-578.

[15] Subhramannia, M., Pillai V. K. *J Mater Chem,* 2008, *18*, 5858-5870.

[16] Gotz, M., Wendt, H. *Electrochim Acta,* 1998, *43*(24), 3637-3644.

[17] Hoogers, G. *Fuel Cell Technology Handbook*, David Thompson D., Florida, CRC Press, LLC, 2003.

[18] Shen, S. Y., Zhao, T. S., Xu, J. B., Li, Y. S. *J Power Sources.* 2010, *195*, 1001-1006.

[19] Ma, L., He, H., Hsu, A., Chen, R. *J Power Sources*, 2013, *241*, 696-702.

[20] Yimin Zhu, Y., Zakia Khan, Z., Masel, R. I. *J Power Sources.* 2005, *139*, 15-20.

[21] Wang, Z. B., Zuo, P. J., Yin G. P. *Fuelcells.* 2009, *9*, 106-113.

[22] Lamy, C., Lima, A., LeRhun, V., Delime, F., Coutanceau, C., Leger, J.-M. J. Power Sources. 2002, 105, 283-296. ["Reprinted with permission from Elsevier]

[23] Tsitomu Ioroi, T., Tomoki Akita, T., Shin-ichi Yamazaki, S.-I., Zyun Simora, Z., Naoko Fujiwara, N., Kazuaki Yasuda, K. *Electrochim Acta.* 2006, *52*, 491-498.

[24] Lu, Y., Tu, J., Gu, C., Xia, X., Wang, X., Mao, S. X. *J Mater Chem.* 2011, *21*(12), 4843-4849.

[25] Zhang, J., Ma, H., Zhang, D., Liu, P., Tian, F., Ding, Y. *Phys Chem Chem Phys.* 2008, *10*(22), 3250-3255.

[26] Geraldes, A. N., Da Silva, D. F., Pino, E. S., Spinace, E. V., Neto, A. O., Linardi, M., Dos Santos, M. *J. Braz. Chem. Soc.* 2014, *25*(5), 831-840.

[27] Antolini, E. *Energ Environ Sci.* 2009, *2*, 915-931.

[28] Liu, C., Liu, M., Li, Q., Xu, Q. *Catalysis.* 2015, *5*, 1068-1078.

[29] Chen, Y., Zhuang, L., Lu J. *Chin. J Catal.* 2007, *28,* 870-874.

[30] Takamura, T., Sato, Y. *Electrochim Acta.* 1974, *19*, 63-68.

[31] Lim, B., Jiang, M., Camargo, P. H. C., Cho, E. C., Tao, J., Lu, X., Zhu, Y., Xia, Y. *Science.* 2009, *324*, 1302-1305.

[32] Liang, Z. X., Zhao, T. S., Xu, J. B., Zhu, L. D. *Electrochim Acta.* 2009, *54*, 2203-2208.

[33] Jurzinsky, T., Cremens, C., Jun, F., Pinkwar, K. *Int J Hydrogen Energ.* 2015, *40*(35), 1-8.

[34] Qi, Z., Geng, H., Wang, X., Zha, C., Ji, H., Zhang, C., Xu, J., Zhang, Z. *J Power Sources,* 2011, *196*, 5823-5828.

[35] Yi, Q., Chu, H., Chen, Q., Yang, Z., Liu X. *Electroanal,* 2015, *27*, 388-397.

[36] Lu, Y., Tu, J., Gu, C., Xia, X., Wang, X., Mao, S. X. *J Mater Chem,* 2011, *21*(12), 4843-4849.

[37] Mani, P., Srivastava, R., Strasser, P. *J Phys Chem,* 2008, *112*, 2770-2778.

[38] Shen, P. K., Xu, C. *Electrochem Commun,* 2006, *8*(1), 184-188.

[39] Ocampo, A. L., Castellanos, R. H., Sebastian, P. T. *J New Mat Elect Syst,* 2002, *5*, 163-168.

[40] Li, Z.-Y., Ye, K.-H., Zhong, Q.-S., Zhang, C.-J., Shi, S.-T., Xu, C.-W. *Chempluschem.* 2014, *79(11),* 1569-1572.

[41] Fu, G., Liu, Z., Chen, Y., Lin, J., Tang, Y., Lu, T. *Nano Research.* 2014, *7*(8), 1205-1214.

[42] Lim, B., Kim, J. W., Hwang, S. J., Yoo, S. J., Cho, A. E., Lim, T.-H., Kim, S.-K. *Bull Korean Chem Soc.* 2010, *31*(6), 1577-1582.

[43] . Xiong, Y., Xia, Y., *Adv. Mater*. 2007, *19*, 3385–3391

[44] Kuai, L., Geng, B., Wang, S., Zhao, Y., Luo, Y., Jiang, H. *Chem A Euro J.* 2011, *17*(12), 3482-3489.

[45] Wang, C., Daimon, H., Onodera, T., Koda, T., Sun, S. *Angew Chem Int Edit.* 2008, *47*, 3588-3599.

[46] Alia, S. M. *Adv Fun Mater.* 2010, *20*, 3742-3746.

[47] Yu, T., Kim, D. Y., Zhang, H., Xia, Y. *Ange Chem Int Edit.* 2011, *50*(12), 2773-2777.

[48] Liang, Z. X., Shi, J. Y., Liao, S. J., Zeng, J. H. *Int J Hydrogen Energ.* 2010, *35*, 9182-9185.

[49] Koh, S., Strasser, P. *J Am Chem Soc Comm.* 2007, *129*, 12624-12625.

[50] Nishanth, K. G., Sridhar, P., Pitchumani, S., Shukla, A. K. *J Electrochem Soc.* 2011, *158*(8), B871-B876.

[51] Lim, B., Jiang, M., Camargo, P. H. C., Cho, E. C., Tao, J., Lu, X., Zhu, Y., Xia, Y. *Science.* 2009, *324*, 1302-1305.

[52] Sarkar, A., Murugan, A. V., Manthiram, A. *J Mater Chem.* 2009, *19*(1), 159-165.

[53] Wang, W., Zheng, D., Du, C., Zou, Z., Zhang, X., Xia, B., Yang, H., Akins, D. L. *J Power Sources.* 2007, *167*, 243-249.

[54] Liu, Z. C., Shen, W. H., Bu, W. B., Chen, H. R., Hua, Z. L., Zhang, L. X., Li, L., Shi, J. L., Tan, S. H. *Micro Meso Mater,* 2005, *82*, 137-145.

[55] Fernandez, J. L., Walsh, D. A., Bard, A. J. *J Am Chem Soc,* 2005, *127*, 357-365.

[56] Wang, B. *J Power Sources.* 2005, *155*, 1-15.

[57] Koh, S., Strasser, P. *J. Am. Chem. Soc Comm.* 2007, *129*, 12624-12625.

[58] Choi, H.-J., Jung, S.-M., Seo, J.-M., Chang, D. W., Dai, L., Baek, J.-B. *Nanoenerg.* 2012, *1*, 534-551.

[59] Bezerra, C., Zhang, L., Marques, A. L. B., Zhang, J. *J Power Sources.* 2007, *173*, 891-908.

[60] Valisi, A. S., Maiyalagan, T., Khotseng, L., Linkov, V., Pasupathi S. *Electrocatal.* 2012, *3*, 108–118.

[61] Takasu, Y., Ohashi, N., Zhang, X. G., Mukarami, Y. *Electrochim. Acta.* 1996, *41*, 2595-2600.

[62] Watanabe, M., Tryk, D. A., Wakisaka, M. Yano, H., Uchida H., *Electrochim. Acta*. 2012, *84*, 187-201.

[63] Higuchi, E., Uchida, H., Watanabe, M. *J. Electroanal. Chem.* 2005, *583,* 69-76.

[64] Shih, Y.-H., Sagar, G. V., Lin, S. D. *J. Phys. Chem C.* 2008, *112*, 123-130.

[65] Shao, M., Peles, A., Shoemaker, K., *Nanoletters.* 2011, *11*, 3714-3719.

[66] Wittkopt, J. A., Zheng, J., Yan, Y. *ACS Catal.* 2014, *4*, 3145-3151.

[67] Ralph, R. T., Hogarth, M. P. *Platinum Metals Rev.* 2002, *46*, 117-135.

[68] Gotz, M., Wendt, H. *Electrochim Acta.* 1998, *43*(24), 3637-3644.

[69] Martin, L. G. Synthesis and characterization of PtSn/C cathode catalyst via polyol reduction method for use in direct methanol fuel cell, thesis, University of the Western Cape, 2013.

[70] Song, C., Zhang, J. PEM fuel cell electrocatalysts and catalyst layers, fundamental and applications. In *Electrocatalytic Oxygen Reduction Reaction*, J. Zhang, Editor. LondonPlace: Springer, pp. x–x. 89-134.

[71] Pronkin, S. N., Bonnefont, A., Ruvinskiy, P. S., Savinova, E. *Electrochim Acta,* 2010, *55*, 3312-3323.

[72] Liu, Z., Hong, L., Tham, M. P., Lim, T. H., Jiang, H. *J Power Sources,* 2006, *161*, 831-835.

[73] Li, W., Zhao, X., Manthiaram, A. *J Mater Chem A.* 2014, *2*, 3468-3476.

[74] Guo, J. W., Zhao, T. S., Prabhuram, J., Chen, R., Wong, C. W. *Electrochim Acta.* 2005, *51*, 754-763.

[75] Stankovich, S., Dikin, D.-A., Dommett, G. H. B., Kohlhaas, K. M., Zimney, E. J., Stach, E. A., Piner, R. D., Nguyen, S. T., Ruoff, R. S., *Letters.* 2006, *442*, 282-286.

[76] Motero-Ocampo, C., Garcia, J. R. V., Estrada, E. A. *Int J Electrochem Sci.* 2013, *8*, 12780-12800.

[77] Xue, X., Liu, C., Xing, W., Lu, T. J *Electrochem Soc.* 2006, *153*(5), E79-E84.

[78] Bensebaa, F., Patrito, N., Page, Y. L., Ecuyer, P. L., Wang, D. *J. Materials Chem.* 2004, *14*, 3378-3384.

[79] Antolini, E. *App Cataly B: Environ.* 2009, *88*, 1-24.

[80] Takasu, Y., Fujiwara, T., Mukarami, Y., Sasaki, K., Oguri, M., Asaki, T., Sugimoto, W. *J Electrochem Soc.* 2000, *147*, 4421-4427.

[81] Dickinson, A. J., Carrette, L. P. L., Collins, J. A., Friedrich, K. A., Stimming, U. *Electrochim Acta.* 2002, *47*, 3733-3739.

[82] Wang, X., Hsing, I. *Electrochim Acta.* 2002, *47*, 2981-2987.

[83] Kim, T., Takahashi, M., Nagai, M., Kobayashi, K. *Electrochim Acta.* 2004, *50*, 813–817.

[84] Xiong, L., Manthiram, A. *Solid Ionics.* 2005, *176(3-4)*, 385-392.

[85] Zhang, X., Chan, K. *Chem Mater.* 2003, *15*, 451-459.

[86] Sharma, S., Pollet, B. G. *J Power Sources.* 2012, *208*, 96-119.

[87] Singh, B., Dempsey, E. *RSC Adv.* 2013, *3*, 2279-2287.

[88] Sevjidsuren, G., Zils, S., Kaserer, S., Wolz, A., Ettinghausen, F., Dixon, D., Schoekel, A., Roth, C., Schoekel, A., Roth, C., Altantsog, P., Sangaa, D., Ganzorig, C. *J Nanomater.* 2010, *2010*, 1-9.

[89] Li, W., Liang, C., Zhou, W., Qiu, J., Zhou, Z. *J Phys Chem B.* 2003, *107*, 6292-6299.

[90] Bianchini, C., Shen, P. K. *Chem Rev.* 2009, *109*, 4183-4206.

[91] Usgaocar, A. R., Harold, M. H., Chong, H. M., De Groot, C. H. *J Nanosci.* 2014. *2014*, 1-7.DOI:10.1155/2014/404519.

[92] Tsai, Y.-C., Hong, Y.-H. *J. Solid State Electrochem.* 2008, *12*, 1293-1299.

[93] Geraldes, A. N., Silva, J. C. M., De Sa, O. A. *J Power Sources.* 2015, *275*, 189-199.

[94] Qi, Z., Geng, H., Wang, X., Zhao, C., Ji, H., Zhang, C., Xu, J., Zhang, Z. *J Power Sources*. 2011, *196*, 5823-5828.

[95] Hu, F., Chen, C., Wang, Z., Wei, G., Shen, P. K. *Electrochim Acta*. 2006, *52*(3), 1087-1091.

[96] Park, S., Shao, Y., Kou, R., Viswanathan, V. V., Towne, S. A., Rieke, P. C., Liu, J., Lin, Y., Wang, Y. *J Electrochem Soc*. 2011, *158*(3), B297-B302.

[97] Tang, J. M., Jensen, K., Waje, M., Li, W. *J. Phys. Chem. C*. 2007, *111*(48), 17901-17904.

[98] Fang, W. C. *Nanoscale Res Lett*. 2010, *5*, 68-73.

[99] Nagaiah, T. C., Kundu, S., Bron, M., Muhler, M., Schuhmann, W. *Electrochem Comm*. 2010, *12*, 338-341.

[100] Tsuji, M., Kubokawa, M., Yano, R., Miyamae, N., Tsuji, T., Jun, M. S., Hong, S., Lim, S., Yoon, S. H., Mochida, I. *Langmuir*. 2007, *23*(2), 387-390.

[101] Zhiong, L., Manthiram, A. *Solid State Ionics*. 2005, *176*, 385-392.

[102] Che, G., Lakshmi, B. B., Martin, C. R., Fisher, E. R. *Langmuir*. 1999, *15*, 750-758.

[103] Wei, J., Zhou, D., Sun, Z., Deng, Y., Xia, Y. *Advanced Functional Materials*. 2013, *23*(18), 2322–2328.

[104] Bessel, C. A., Laubernds, K., Rodriguez N. M., Baker R. T. K., *J. Phys Chem B*. 2001, *105*, 1115-1118.

[105] Jeng, K.-T., Chien, C.-C., Hsu, N.-Y., Yen, S.-C., Chiou, S.-D., Lin, S.-H., Huang, W.-M. *J Power Sources*. 2006, *160*, 97-104.

[106] Yoo, E., Okata, T., Akita, T., Kohyama, M., Nkamura, J., Honna, I. *Nano Lett*. 2009, *9*(6), 2255–2259.

[107] Wang, S., Jiang, S. P., Wang, X. *Electrochim Acta*. 2011, *56*, 3338-3344.

[108] Jafri, R. I., Rajalakshmi, N., Ramaprabhu, S. *J Mat Chem,* 2010, *20*, 7114-7117.

[109] An, G., Yu, P., Mao, L., Sun, Z., Liu, Z., Miao, S., Miao, Z., Ding, K. *Carbon*. 2007, *45*, 536-542.

[110] Ha, H. W., Kim, I. Y., Hwang, S. J., Ruoff, R. S. *Electrochem Solid State Letters*. 2011, *14*, B70-B73.

[111] Yaldagard, M., Jahanshahi, M., Seghatoleslami, N. *World J Nanosci and Eng.* 2013, *3*, 121-153.

[112] Ismagilov, Z., Kerzhentsev, M., Shikina, N., Lisitsyn, A., Okhlopkova, L., Barnakov, C. N., Sakashita, M., Iijima, T., Tadokoro, K. *Catal Today.* 2005, *102-103*, 58-66.

[113] Salgado, J. R. C., Alcaide, F., Alvarez, G., Calvillo, L., Lazaro, M. J., Pastor, E. *J. Power Sources.* 2010, *195*, 4022-4029.

[114] Ding, J., Chan, K. Y., Ren, J., Xiao, F. *Electrochim Acta.* 2005, *50*, 3131-3141.

[115] Liu, R., Wu, D., Feng, X., Mullen, K. *Angew. Chem Int. Ed.* 2010, *49*, 2565-2569.

[116] Sebastian, D., Lazaro, M., Suelves, I., Moliner, R., Baglio, V., Stassi, A., Arico, A. *Int J Hydrogen Energ.* 2011, *37*(7), 6253-6260.

[117] Huerta, R. G., Vulenzuela, M. A., Garcia, R. V., Alonso-Vante, N., Velazquez, M. T., Ruiz-Camaelo, B. *J New Mat Electroc Syst.* 2012, *15*(3), 123-128.

[118] Hyeon, T., Han, S., Sung, Y. E., Park, K. W., Kim, Y. W. *Ange Chem Int Edit.* 2003, *115*(36), 4488-4492.

[119] Yuan, W., Scott, K., Cheng, H. *J Power Sources.* 2006, *163*, 323-329.

[120] Stankovich, S., Dikin, D.-A., Dommett, G. H. B., Kohlhaas, K. M., Zimney, E. J., Stach, E. A., Piner, R. D., Nguyen, S. T., Ruoff, R. S. *Letters.* 2006, *442*, 282-286.

[121] Li, Z.-Y., Ye, K.-H., Zhong, Q.-S., Zhang, C.-J., Shi, S.-T., Xu, C.-W. *Chempluschem.* 2014, *79*(11), 1569-1572.

[122] Du, S. *J Power Sources.* 2010, *195*, 289-292.

In: Direct Methanol Fuel Cells
Editors: Rose Hernandez et al.
ISBN: 978-1-53612-603-7

Chapter 2

DMFC WITH A FUEL TRANSPORT LAYER USING A POROUS CARBON PLATE FOR NEAT METHANOL USE

Nobuyoshi Nakagawa[1,*], Mohammad Ali Abdelkareem[2,3], Takuya Tsujiguchi[4] and Mohd Shahbudin Masdar[5]

[1]Division of Environmental Engineering Science,
Gunma University, Kiryu, Gunma, Japan
[2]Department of Sustainable and Renewable Energy Engineering,
University of Sharjah, Sharjah, UAE
[3] Department of Chemical Engineering,
Minia University, Elminia, Egypt
[4]Faculty of Mechanical Engineering, Kanazawa University,
Kanazawa, Ishikawa, Japan
[5]Fuel Cell Institute, Universiti Kebangsaan Malaysia,
Selangor, Malaysia

ABSTRACT

In direct methanol fuel cells, methanol crossover (MCO) through the electrolyte membrane is a major problem to be solved. Due to the MCO, degradation of the power generation and energy loss occur, hence a diluted methanol solution with 1 to 3 M (3 to 10 wt.%) is generally used as the fuel, which limits achieving a high energy density that is an attractive feature of the DMFC. The authors have proposed a fuel supply layer using a porous carbon plate (PCP) for DMFCs that allowed using high methanol concentrations up to neat methanol. The layer consists of a thin PCP and a gap layer that has a mechanism to supply the methanol as a vapor. This effective and practical technique is suitable for the utilization of a high methanol concentration and miniaturization of the DMFC system. In this chapter, we described the principle and practical MCO control at the fuel supply layer using the PCP. The power generation performance of the DMFCs equipped with the fuel supply layer, the mass transport characteristics through the membrane, and the development of prototype stacks are explained. The pore structure and wettability of the PCP were found to have a significant impact on the MCO. Since the fuel supply is conducted in the vapor phase, water transport through the membrane was quite different from that by the liquid feed, as well as affecting the cathode performance and reaction products. For practical applications, prototypes of the DMFC's stack with the PCP were designed and tested. The power generation profiles of the passive and active stacks are then discussed in detail in this chapter.

Keywords: methanol crossover, fuel transport layer, porous carbon plate, neat methanol use, vapor feed, DMFC stack

1. INTRODUCTION

The exponential growth of the applications of portable devices during the last two decades requires the development of new energy devices that have much higher energy densities to meet their expanded power requirements. The current battery technologies can no longer meet the energy demand of these devices, therefore, scientists are developing new power systems [1] that can meet the power requirements of these devices.

Direct methanol fuel cells (DMFCs) are the best candidate for this purpose due to their high theoretical energy density, 4800 Wh L^{-1}, that can satisfy the power requirements of the current and next generation portable devices. Unfortunately, the actual energy density of the current DMFCs is too low to be used in powering these portable devices due to the methanol crossover (MCO) and the sluggish electrode reactions [2-5]. Although tremendous work has been done to eliminate or reduce the MCO whether by modifying the current Nafion membrane or producing new ones [6-10], the Nafion membrane is still the best one for the DMFCs.

Due to the MCO, the methanol concentration used in DMFCs is limited; the optimum ones ranged from 1 to 3 M (3 to 10 wt.%) [11, 12], thereby, the actual energy density is within one tenth of its theoretical value. The use of a high methanol concentration, the target being 100% methanol, is one of the best ways to increase the energy density of the DMFC. Some reports considered that the control of the mass transport in the backing layer or by adding a porous material in front of the anode surface is an effective way to use high methanol concentrations [14-16].

The authors proposed a fuel supply layer using a porous carbon plate (PCP) [17-26] that controlled the methanol mass transport rate, therefore, high methanol concentrations, up to neat methanol, could be effectively used. The layer has a simple structure, i.e., just placing a thin porous plate between the fuel reservoir and the anode current collector. The important factor for the success of using a high concentration methanol is feeding methanol in the vapor phase through the transport layer. The methanol flux depends on the properties of the porous plate, i.e., porosity, pore size, thickness, water absorptivity, etc., and the properties of the layer, i.e., pressure, thickness.

In this chapter, we provide a summary of our studies, conducted during the past ten years, aimed at developing a DMFC operating with neat methanol. It includes mass transport through the transport layer by the PCP and performance of DMFCs with PCP, from a basic study using a horizontally-arranged passive DMFC to prototypes of the passive and active DMFC stacks.

2. Studies Using a Horizontally-Oriented Passive DMFC with PCP

In this section, fundamental studies of the mass transport control though the fuel supply layer using a PCP having different pore structures conducted with a passive horizontally-oriented DMFC were introduced.

2.1. Fuel Supply Layer Using a PCP

By placing the PCP on the surface of the membrane electrode assembly (MEA), the permeation rate of methanol and water through the fuel supply layer was controlled depending on the physical transport resistance through the porous media. The resistivity of the PCP depends on its properties, such as porosity, thickness, pore size, water absorptivity, etc., as well as the operating conditions, i.e., open circuit and closed circuit of the DMFC. The authors investigated the permeation rates of methanol and water through the fuel supply layer using PCPs with different properties [18].

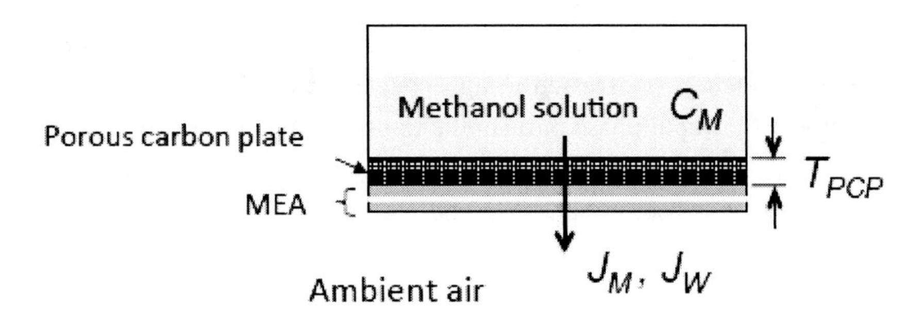

Figure 2-1. Schematic diagram of the fuel transport layer by placing the PCP on the anode of the MEA of a horizontally-oriented passive DMFC.

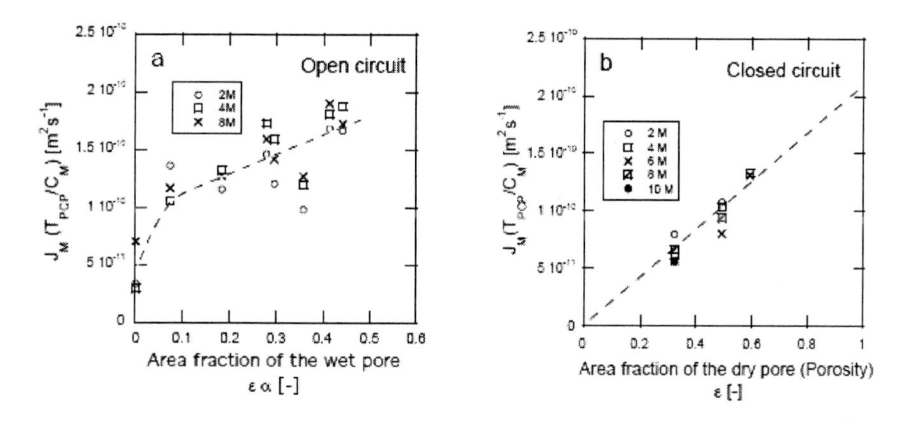

Figure 2-2. The diffusivity of the methanol transport through the PCP, (a) a function of the fraction of the wet pore area under open circuit conditions, and (b) a function of the fraction of the dry pore area (porosity) under closed circuit conditions.

In the open circuit condition, the aqueous methanol solution permeates through the wet pores of the PCP to the MEA. Thereby, the methanol flux, J_M, through the layer and an effective diffusivity, $J_M\ (T_{PCP}/C_M)$, where T_{PCP} is the thickness of the PCP and C_M is the methanol concentration, became a function of the fraction of the wet pore area, $\varepsilon{\cdot}\alpha_w$, of the PCP as shown in Figure 2-2a, where ε is the porosity and α_w is the water absorptivity that is defined by the ratio of the volume of water absorbed to the volume of the open pores. The diffusivity increased with the increasing water absorptivity suggesting that the methanol flux dominated through the wet pores. The non-zero and positive diffusivity at $\varepsilon{\cdot}\alpha_w$=0 means contribution of the flux through the dry open pores.

On the other hand, under the closed circuit condition, the diffusivity proportionally increased with the increase in the porosity ε as shown in Figure 2-2b. When the circuit was closed, CO_2 gas was produced at the anode by the electrode reaction. The CO_2 first filled the open pores of the PCP, then it exits the PCP through the pores. At this time, the methanol has to be transported through the open pores as a vapor. The effective diffusivity of the methanol through the layer is then proportional to the porosity of the PCP. The fuel supply layer using a PCP converts the liquid methanol feed to the vapor methanol feed. The porosity of the PCP is one of the main factors in controlling the MCO [18].

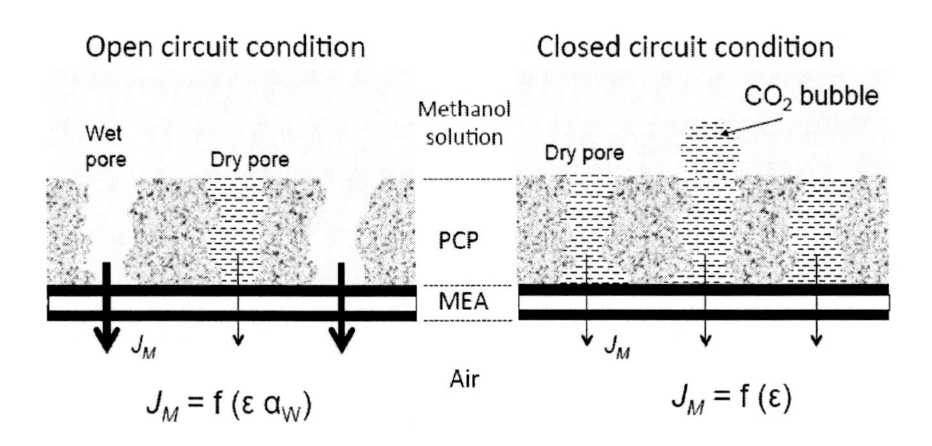

Figure 2-3. The mechanism of the methanol transport through the PCP differed between the open circuit conditions and closed circuit conditions.

When the concentration of methanol is 40wt.% or higher, the surface tension of the methanol solution becomes quite low, less than 30 mN m^{-1}, and the solution easily wets and spreads over the pore surface. This may result in the direct contact of the highly concentrated methanol solution to the MEA, thereby, irreversible damage of the membrane [27]. To avoid this situation, we inserted a gap space by placing a plate with a large open ratio between the PCP and MEA. The gap space, a gas layer, ensures the vapor feed of the methanol solution through the fuel supply layer is formed and prevents the direct contact between the highly concentrated methanol and the MEA. Hence, based on the above mechanism and results, we have proposed the fuel supply layer shown in Figure 2-4 in order to use the highly concentrated methanol in the DMFC [19]. The fuel supply layer is a simple layer structure with a PCP placed on the anode side of the MEA with a gap space, i.e., a gas layer, between the PCP and MEA.

The fuel supply layer has the function of converting the methanol solution into methanol vapor and then supplying the methanol vapor to the anode of the MEA. Therefore, it is a vapor feed DMFC. The PCP provides an air/water interface on the pore inside and/or the outer surface of the PCP, and vaporization of the methanol occurs in the gas layer. When the methanol oxidation occurs at the anode, the CO_2 gas product accumulates in the gas layer, then flows outside through the pores of the PCP. The direction of the methanol transport is counter to that of the CO_2 gas inside the pores. The strong hydrophobicity of the PCP and the CO_2 counter flow through the pores interferes with the intrusion of the methanol into the gas layer. The methanol flux through the supply layer strongly depends on the properties of the PCP and the gas layer, i.e., thickness, pore size, porosity, hydrophobicity, etc., related to the area and position of the air/water interface where the vaporization occurs and a resistance flow through the PCP.

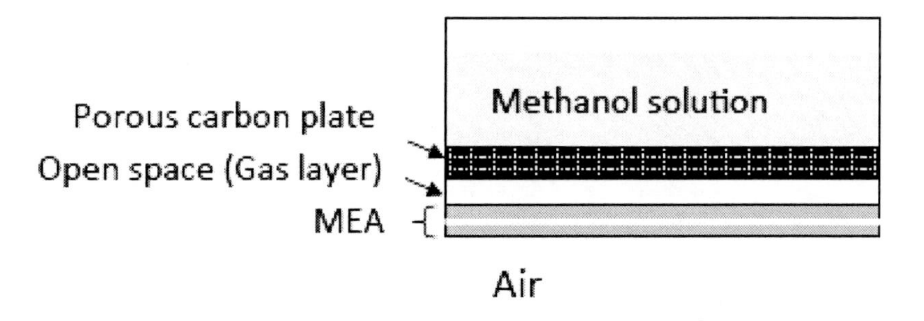

Figure 2-4. Structure of the fuel supply layer using the PCP for the highly concentrated methanol use.

The fuel supply layer using the PCP is a very simple layered structure with a limited volume expansion. Although some other technologies of the methanol vapor feed have been proposed [28], they usually need auxiliary equipment or external power [29-33]. The porous media we used was made of carbon that is advantageous due to its high durability against chemical corrosion and high electric conductivity. The materials are not limited to carbon. We have confirmed that a porous ceramic plate [18] and a perforated metal sheet [34] can also be effectively used. Followed by our first report on the DMFC using the PCP [19], for the highly concentrated methanol use, a similar fuel transport layer using a pervaporation membrane [16, 35-40] or sintered metal plate [41-42] instead of the PCP were reported by other researchers.

2.2. Performance of a Passive DMFC Using a Horizontally-Oriented PCP

The power generation and the methanol and water transports through the MEA were investigated for the DMFCs with the fuel supply layer using PCP [19]. Using a DMFC illustrated in Figure 2-5, the cell performance was investigated with and without the PCP at different methanol concentrations.

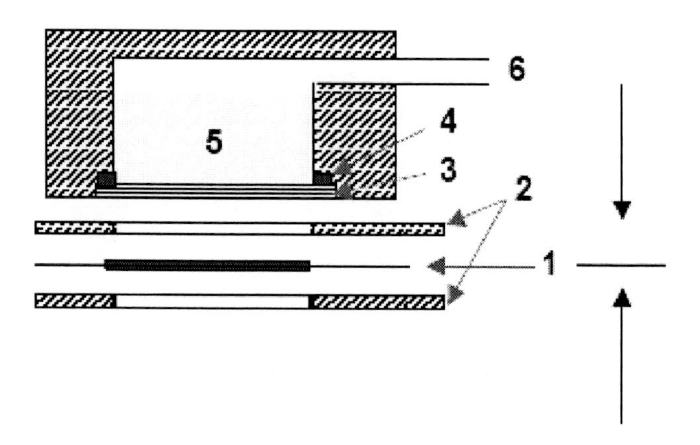

1. MEA
2. Current collectors
3. Porous carbon plate, PCP
4. Rubber sheet
5. Methanol solution
6. Open tube

Figure 2-5. Schematic diagram of a passive DMFC with a fuel transport layer using the PCP. Reproduced with permission from [18]. Copyright © 2007 Elsevier.

The DMFC used in these experiments was passive type and horizontally oriented. As shown in Figure 2-5, the cell had a methanol reservoir of 12 cm^3 (22-mm wide, 22-mm long and 25-mm depth). The MEA was sandwiched between two current collectors, which were gold-plated stainless steel plates of 2-mm thickness with open holes. The holes of the anode current collector, with an open ratio of 73%, provided the gas layer between the PCP and the MEA. The MEA consisted of Nafion 112 and two carbon clothes with a catalyst layer. Pt black and PtRu black were used for the cathode and anode, respectively. The projected area of the electrode was 5 cm^2. The PCP shown in Table 2-1 was placed on the anode current collector. The horizontal arrangement resulted in a homogeneous distribution of the fuel over the transport layer and made the analysis of the experimental results easier.

A comparison of the power generation characteristics between the DMFC with the PCP and that without the PCP using the different methanol concentrations is shown in Figure 2-6, i.e., current-voltage (i-V) curves, (a) and (b), and current-time (i-t) curves, (c) and (d). The PCP used in these measurements was Y2 with the properties listed in Table 2-1.

Table 2-1. Properties of the porous carbon plates used for the fuel supply layer. Reproduced with permission from [18]. Copyright © 2007 Elsevier

PCP	Thickness [mm]	Water absorptivity α_w [-]	By the mercury porosimeter			By the perm-porometer	
			Pore volume [cm^3 g^{-1}]	Average pore diameter [μm]	Porosity ε [-]	Bubble point pressure [kPa]	Bubble point diameter [μm]
Y1	1.0	0.40	0.543	42.3	0.417	3.05	14.8
Y2	2.0	0.21					
S1	1.0	0.15	0.556	1.43	0.457	42.8	1.05

The i-V curves of the conventional DMFC without the PCP exhibited relatively high open circuit voltages and the best i-V curve at 2 M showed the highest cell voltage over 50 mA cm^{-2} in Figure 2-6a. The cell voltage decreased with a further increase in the methanol concentration. The voltage decrease was mainly due to the methanol crossover which was confirmed by the fluctuating current densities at the higher methanol concentrations, low cell voltages, and directly by the i-t curves shown in Figure 2-6c. The current density at the constant cell voltage was initially high, but rapidly decreased to less than one-third of its initial value within 1h, then it further decreased with time. This was related to the severe flooding of the cathode by the accumulated water and water transported from the anode to the cathode side.

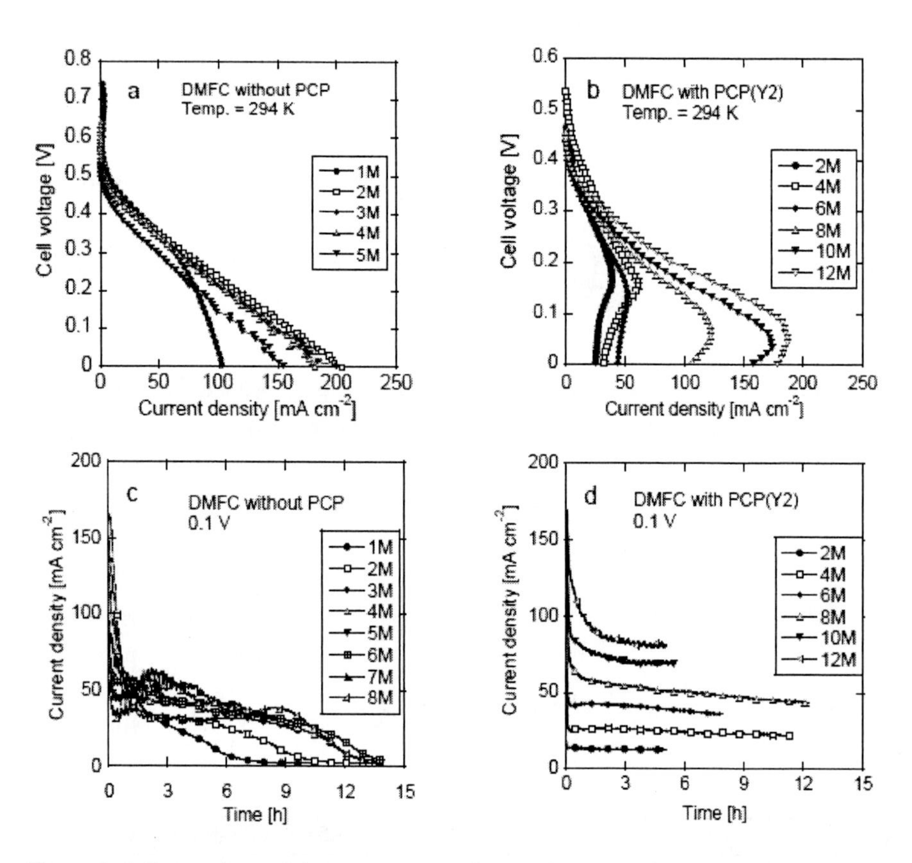

Figure 2-6. Comparison of the power generation performance, i.e., i-V curves, (a) and (b), and current density versus time, (c) and (d), between passive DMFCs with and without the PCP. Reproduced with permission from [16]. Copyright © 2006 Elsevier.

On the other hand, for the DMFC with PCP(Y2), although the open circuit voltages were slightly lower than those of the DMFC without the PCP, the i-V performance increased with the increasing methanol concentration up to 12 M. In this case, the current density at the low cell voltages, less than 0.1 V, clearly showed a limiting current, caused by the restriction of the methanol supply rate. The limiting current was also confirmed by the steady currents observed in the i-t curves (Figure 2-6d) that were proportional to the methanol concentration. The high stability of the current density with time for the DMFC with the PCP suggested that the cathode was almost free from water flooding. Such a significant difference in the DMFC performance was due to the change in the mass transport of methanol and water through the MEA by the fuel transport layer with the PCP.

The steady current density of the DMFC with the PCP(Y2) that was proportional to the methanol concentration was re-plotted in Figure 2-7a along with adding data at the higher methanol concentrations up to 100% (24.7 M). The straight relationship started from the origin to 20 M suggested that the DMFC operated under the rate limiting step of the methanol supply in the concentration range. The slope of the straight line corresponds to the diffusivity of the methanol to the anode through the MEA. The smaller, about a half, slope of the straight line for the DMFC with the PCP than that for the DMFC without the PCP at the lower methanol concentrations reflected the strongly controlled methanol transport, about half the diffusivity, by the fuel transport layer. The current densities with the decreased values that deviated from the straight line were due to the uncontrolled MCO and another rate limiting step.

The methanol and water fluxes through the MEA were measured by the weight loss and the methanol concentration of the fuel in the reservoir as well as the current density during the power generation at a constant cell voltage. The comparison of the methanol and water fluxes between the DMFC with and without the PCP(Y2) is shown in Figure 2-7b. The methanol flux of the DMFC with the PCP was significantly decreased, down to about 1/10 compared to that without the PCP. The MCO at 20 M for the DMFC with PCP was nearly equivalent to that at 7 M for that without the PCP. At the same time, the water flux was also significantly changed (Figure 2-7c). The water flux was 0.1 g m^{-2} s^{-1} regardless of the methanol concentration in the case without the PCP. The positive value indicated that the direction of the water flux was from the anode to the cathode through the MEA. Whereas, in case of using the PCP, the water flux decreased with increasing methanol concentration, and, noteworthy, it became negative from a certain concentration, 5 M, and further negatively increased as the methanol concentration increased. The negative flux indicated back diffusion from the cathode to the anode. With the increasing methanol concentration, the shortage of water for the anodic reaction became serious. The water produced at the cathode diffused to the anode through the membrane based on the water concentration difference between the anode and the cathode. The water back diffusion from the cathode is advantageous in preventing the cathode flooding [43]. However, the water supply from the cathode sometimes limits the reaction rate at the anode [22]. Hence, water management at the cathode becomes important in order to activate the electrode reaction and increase the power density of the vapor feed type of the DMFC operating with a high methanol concentration [36, 44-46].

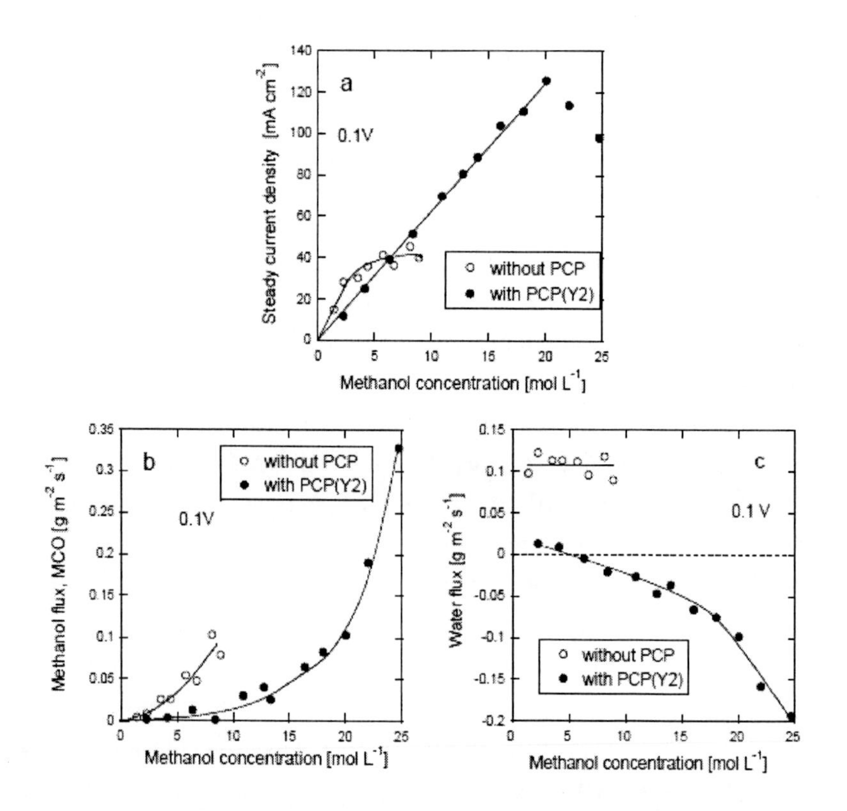

Figure 2-7. Steady current density, (a), methanol flux, (b), and water flux, (c), of the various methanol concentrations for the DMFC with and without PCP. Reproduced with permission from [16]. Copyright © 2006 Elsevier.

2.3. Effect of the PCP Properties on the Cell Performance

The ability of the fuel supply layer using the PCP to control the methanol supply depends on the properties of the PCP and those of the gas layer, i.e., porosity, thickness, pore diameter, gas barrier thickness, pressure, etc. The lower porosity and the greater thickness of the plate (or layer) result in a higher resistance to mass transport through the layer and thus, a lower MCO.

Figure 2-8 shows the power generation of the DMFCs using different PCPs, i.e., Y1, Y2 and S1, shown in Table 2-1. Y1 and Y2 were made of the same amorphous carbon and have a similar pore structure with a 0.543 $cm^3\ g^{-1}$ pore volume, 42 μm average pore diameter and 0.417 porosity, but different thickness, i.e., 1 mm and 2 mm for Y1 and Y2, respectively. S1 was made of graphitic carbon and had a similar pore volume, 0.556 $cm^3\ g^{-1}$, and porosity, 0.457, to that of Y1 and Y2, but a smaller pore diameter, 1.4 μm. The resistivity of the air flow through the plate for S1 was greater than that of Y1 and Y2 due to the smaller pore diameter. The effective diffusivity through the fuel transport layer, the slope of the straight relationship in Figure 2-8a, was the lowest for Y2, while it was slightly higher for S1. These PCPs showed similar MCOs at the different methanol concentrations up to the neat methanol (Figure 2-8b). On the other hand, for Y1, the current density did not show a straight relationship in the high concentration range over 10 M even though the current densities were the highest among the different PSPs. The MCO for Y1 was significantly increased at the high concentrations (Figure 2-8b). The uncontrolled MCO for Y1 was confirmed by the abnormal increase in the cell temperature (Figure 2-8c). For the neat methanol use, the small pore diameter under 10 μm was appropriate for the selected porosity and thickness.

In order to ensure the effectiveness of the fuel supply layer in preventing the liquid permeation into the gas layer through the PCP, it is recommend to increase the pressure of the gas layer to some extent [22]. To that end, the resistivity of the PCP should be kept high enough to avoid any intrusion of the liquid into the gas layer through the pores of the PCP. An increase in the flow-through resistance of the PCP can be realized by increasing the thickness and/or decreasing the pore size. Y2 had a thickness double that of Y1. Y2 thus had a greater flow-through resistance than that of Y1. The average pore diameter of S1 was about 1/30 that of Y1 and Y2. Hence, Y2 and S1 had a higher flow-through resistance than that of Y1 and made possible the use of 100% methanol.

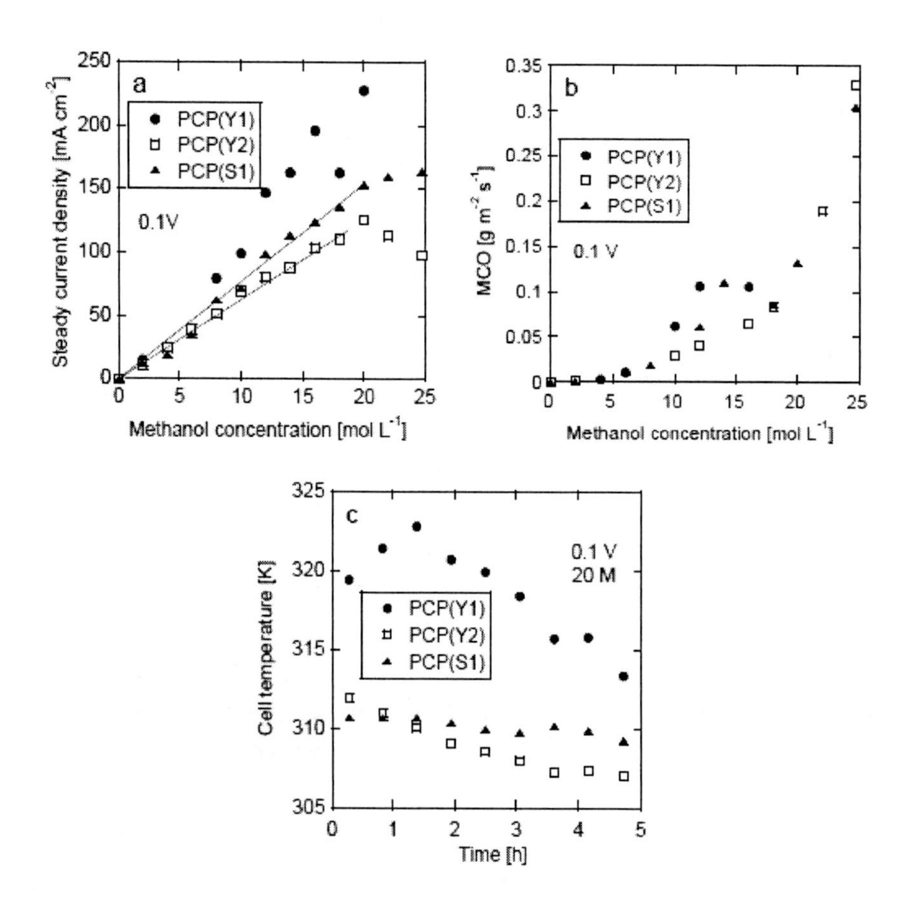

Figure 2-8. Current density, (a), methanol flux (MCO), (b), and the temperature, (c), at the DMFCs with different PCPs. Reproduced with permission from [18]. Copyright © 2007 Elsevier.

Another parameter that affects the methanol transport resistance is the thickness of the gas layer. The increase in the gas layer thickness increased the methanol transport resistance as shown in Figure 2-9 as seen from the decreased slope of the straight line with the increasing gas barrier thickness.

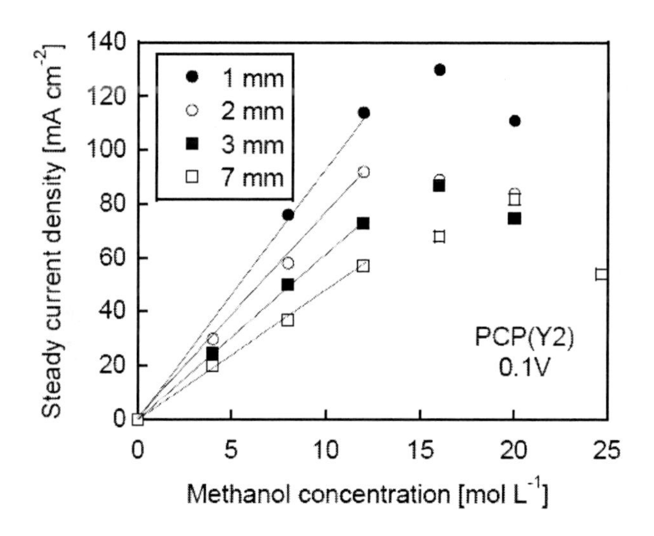

Figure 2-9. Effect of the gas layer thickness on the current density of the DMFC with PCP(Y2). Reproduced with permission from [18]. Copyright © 2007 Elsevier.

Although the flow through resistance of the PCP is not identical to the methanol transport resistance through the PCP, they are related to each other because both of them depend on the pore structure of the PCP. If the flow-through resistance of the PCP is too high, the methanol transport rate through the PCP is reduced and the current density is also reduced. Therefore, it is important for the structure of the PCP and the gas layer to be optimized to maximize the DMFC performance for the methanol concentration to be used.

The stable power generation of the DMFC using highly concentrated methanol solutions significantly increases the energy density of the cell. The volumetric energy density of the DMFC calculated based on the volume of the consumed methanol solution remarkably increased using the PCP as shown in Figure 2-10. The increased energy density of the DMFC with the PCP was greater than seven times that without the PCP. This was related to using a higher methanol concentration, and a higher coulombic efficiency obtained by the controlled MCO.

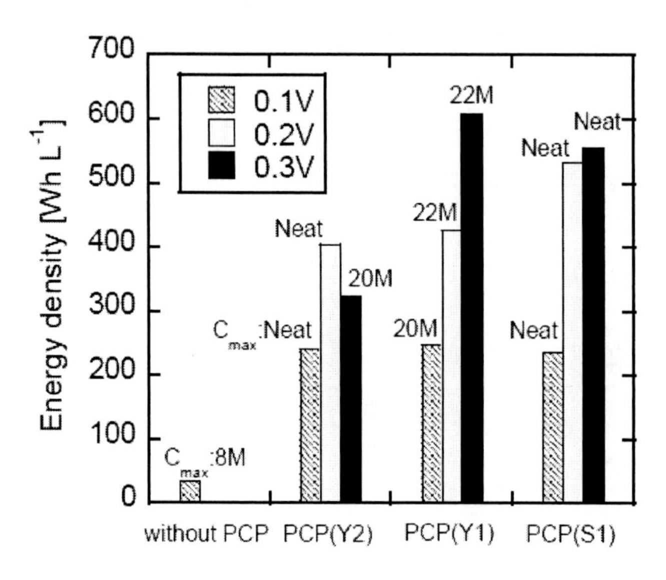

Figure 2-10. Comparison of the energy density of the passive DMFCs with and without the PCP. Reproduced with permission from [18]. Copyright © 2007 Elsevier.

2.4. Improvement of Water Management Using a Hydrophobic Air Filter

In the previous section, it was pointed out that water flooding scarcely occurs at the cathode of the DMFC with the PCP. This is because the water transports from the cathode to the anode, thereby water is less likely to accumulate in the cathode. This especially happens at the high concentrations near 100% which results in a water deficiency at the anode and dehydration of the membrane is likely to occur. These reduced the reaction rate and the power output especially at a low air humidity. Since the methanol oxidation reaction, $CH_3OH + 3/2\ O_2 \rightarrow CO_2 + 2H_2O$, is water generation, the water demand at the anode can be compensated by the water transported from the cathode. In order to enhance the water transport from the cathode to the anode, a hydrophobic air filter (HAF) is an effective way to improve the water management and the cell performance [25]. The HAF, like a compressed sheet of unwoven fabric with carbon black and PTFE, could significantly improve the current density at 30 and

60% relative humidity. The HFA should be placed at a distance, like 6 mm, from the cathode surface, and not placed in contact with the cathode in order to get higher power outputs [25].

2.5. Reaction Products at the Anode

In the conventional liquid feed DMFCs, CO_2 is the main anode product at the anode [47] and small amounts of formaldehyde (HCHOH), methylformate ($HCOOCH_3$), formic acid (HCOOH) and CO are formed [48]. The distribution of the products and the production rate were affected by the water/methanol ratio, temperature, current density and catalyst morphology [49], and methylformate and methylal were preferentially formed when the water/methanol ratio was low, and the percentage of these products was sometimes greater than 60% [49]. Hence, the distribution of the DMFC products operated at a high methanol concentration needs to be analyzed, because it affects the energy conversion efficiency and the energy density of the DMFC.

We have measured the product distribution at the anode of the DMFC with the PCP at the high methanol concentrations [50]. The products, apart from CO_2, were formaldehyde, methylformate and a very small amount of formic acid. The production rates were low and negligible based on the energy efficiency calculation. The production rates of the intermediates could be summarized as a function of the methanol flux through the layer irrespective of the methanol concentration in the reservoir as shown in Figure 2-11b. This was reasonable because the methanol solution was separated from the anode of the MEA by the fuel transport layer and was not in direct contact with the anode surface. The relationship between the production rate of each component and MCO were very similar for the DMFCs with and without the PCP. For both the formaldehyde and formic acid, the production rate was relatively low and slightly dependent on the methanol flux. On the other hand, the production rate of methylformate notably increased with the increasing MCO, and the production rate for the DMFC with the PCP was 10 times higher than that of the DMFC without the PCP. This was due to the lower water/methanol ratio at the anode surface of the DMFC when using the PCP. In case of the DMFC with the PCP at a high methanol concentration, water is limited at the anode and water is transported from the cathode to the anode through the membrane. Although there was such an increased production rate of methylformate, the selectivity of the total intermediate products from the converted methanol was calculated to be at most 1.0%. The energy loss by the intermediate products for the DMFC with the PCP was then negligibly low.

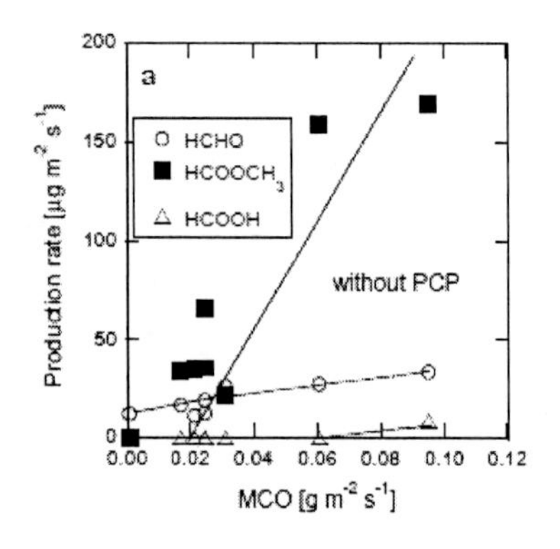

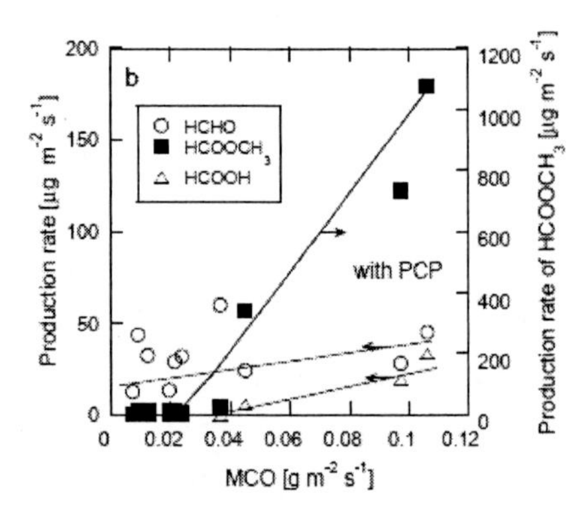

Figure 2-11. The production rate of the intermediate products for the DMFC without the PCP, (a), and with the PCP, (b), operating at different methanol concentrations. Reproduced with permission from [50]. Copyright © 2009 Elsevier.

The gas composition in the anode gas layer was also directly analyzed using a mass spectrometer with a capillary probe [51, 52]. CO_2, methanol and water were found to be the main components and the composition was strongly dependent on the PCP used and the methanol concentration in the reservoir. The vapor pressure of each component, P_{CO2}, P_{CH3OH} and P_{H2O}, were related to the current density. The current density linearly and identically increased with the increase of the P_{CH3OH} in the gas layer up to a certain value irrespective of the type of the PCP used suggesting that the limiting current is due to the methanol transport to the anode. Beyond this value, the current density decreased with the increasing P_{CH3OH} due to the uncontrolled MCO. The calculated liquid methanol concentration equivalent to the CO_2-CH_3OH-H_2O gas in the gas layer was about 5-7 M suggesting that the actual methanol activity at the anode of the DMFC with the PCP was similar to those of the usual liquid feed DMFC even when a very high concentration of methanol was used.

3. Development of a Passive DMFC Stack for Neat Methanol Use

For the practical application of portable and mobile DMFCs, the cell could be operated in different cell orientations, i.e., horizontally or vertically. The previous section described the performance of the horizontally-oriented passive DMFC. In this section, we aimed to extend the methanol transport layer for the vertical arrangement, and designed a passive DMFC stack as well as a demonstration of power generation of the stack using 100% methanol.

So far, some researchers have studied the effect of cell the orientation on the performance of a passive DMFC and they found that vertically-oriented cells produced a better performance than that of the horizontally-oriented ones at moderate and high current densities. The improved performance was related to the increased operating temperature as a result of the higher MCO that in the same time reduced the fuel efficiency [53, 54]. Others showed that the vertical orientation produced a lower performance than the horizontal one, and they related this to the large amount of water that flowed down along the surface of the cathode in the vertical case [55]. These studies revealed that the uncontrollable MCO was the main reason for these variations. As described in the previous section, the DMFC with the fuel transport layer can control the methanol flux through the MEA and prevent the flooding at the cathode. Hence, the DMFC with the fuel transport layer is expected to be beneficial for the vertical cell arrangement.

3.1. Vertically-Arranged Single Cell with PCP

A PCP that functioned as a methanol barrier in the horizontal orientation does not necessarily play a similar role in the vertical orientation, because, in the vertical orientation, there is a non-uniform distribution of the hydrostatic pressure exerted by the methanol solution along with height of the PCP, which has to be taken into consideration for the mechanism of the methanol barrier of the fuel supply layer. The authors investigated the operation of the passive DMFC with PCP that was vertically oriented using different methanol concentrations in order to clarify the proper properties of the PCP that was suitable for the vertical operation [23]. The control of the methanol transport in the vertical operation of the DMFC with the PCP was more sensitive to the properties of the PCP, i.e., bubble point pressure and flow-through resistance, than those in the horizontal orientation. When the bubble point pressure and the flow-through resistance of the PCP were not high enough in the vertical orientation, the hydrostatic pressure of the methanol solution causes a convective liquid flow through the PCP, therefore, the cell performance was affected by the cell orientation. The PCP that had a high pressure CO_2 gas layer, i.e., a high bubble point pressure and a high flow-through resistance, could resist the variations in the hydrostatic pressure [23].

Based on these investigations, we designed a vertically-arranged single cell of a DMFC with a PCP by optimizing the pore structure of the PCP for 100% methanol use (Figure 3-1). The cell had a methanol reservoir with a 12-cm^3 capacity and 55-mm height in the anode compartment. The MEA was sandwiched between two current collectors, which were made of Au coated-stainless steel with open holes having a 74% open ratio, for the transport of the fuel and the oxidant. The thicknesses of the current collectors were 2 mm for the cathode and 1mm for the anode. The PCP with the appropriate pore structure, i.e., a high bubble point pressure (i.e., small pore diameter) and a high flow-through resistance, was placed between the anode current collector and the methanol reservoir. The PCP had a 0.5-mm thickness, 55-mm height and 15-mm width.

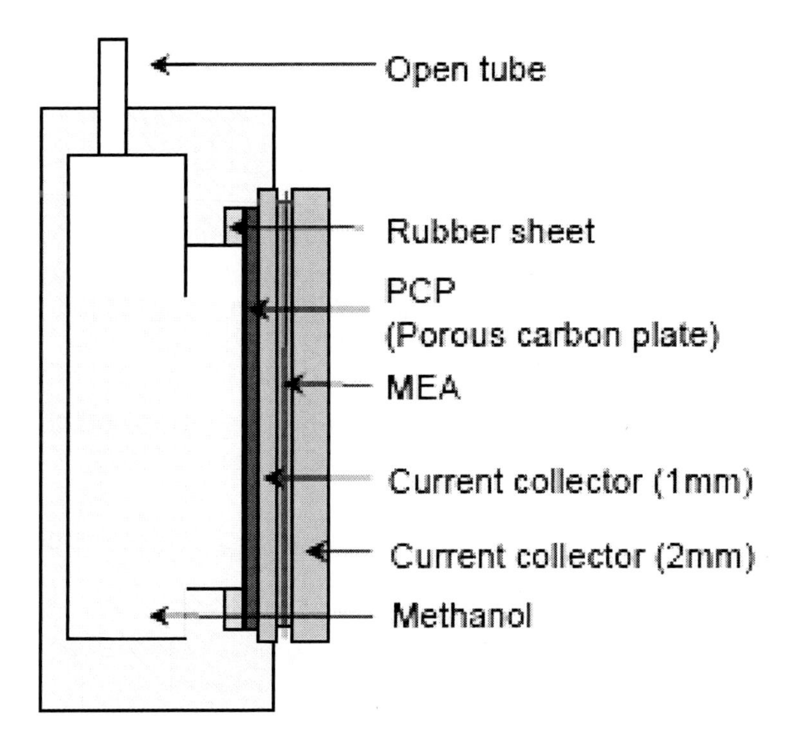

Figure 3-1. Schematic diagram of a vertically-arranged single cell of a DMFC with a PCP. (PCP: 55-mm height, 0.5-mm thickness and 15-mm width) Reproduced with permission from [21]. Copyright © 2012 Elsevier.

Figure 3-2 shows the influence of the methanol head height in the fuel reservoir and cell orientation, i.e., vertical and horizontal, on the power density and cell temperature at 0.25 V using 100% methanol. It was confirmed that the vertical operation with 100% methanol was successfully conducted using the fuel transport layer with the designed PCP. At the beginning of the operation, the height of the methanol was initially 5 mm. The height remained for one hour, then it was increased to 27 mm by adding more methanol. After a one-hour operation, it was further increased to a 55-mm height at which the fuel reservoir was totally filled with methanol. The head was then decreased to 27 mm, then to 5 mm by removing some of the methanol from the reservoir. As shown in Figure 3-2, the power output was about 30 mW cm^{-2} for the 7-hour operation regardless of the head of the methanol solution. This indicated that the PCP used in this experiment had the proper pore structure for sucking methanol to the top of the PCP. However, it should be pointed out that the power density sharply increased to 35 mWcm^{-2} for 1 hour, when the methanol level was changed from 5 mm (span (i)) to 27 mm (span (ii)). This would be related to the increase in the cell temperature caused by the increase in the MCO during this period (span (ii)). The gradual increase in the power density and the cell temperature at span (ii) would be related to the slow sucking rate of the methanol solution into the PCP, based on the capillary action, due to its small pore diameter [23].

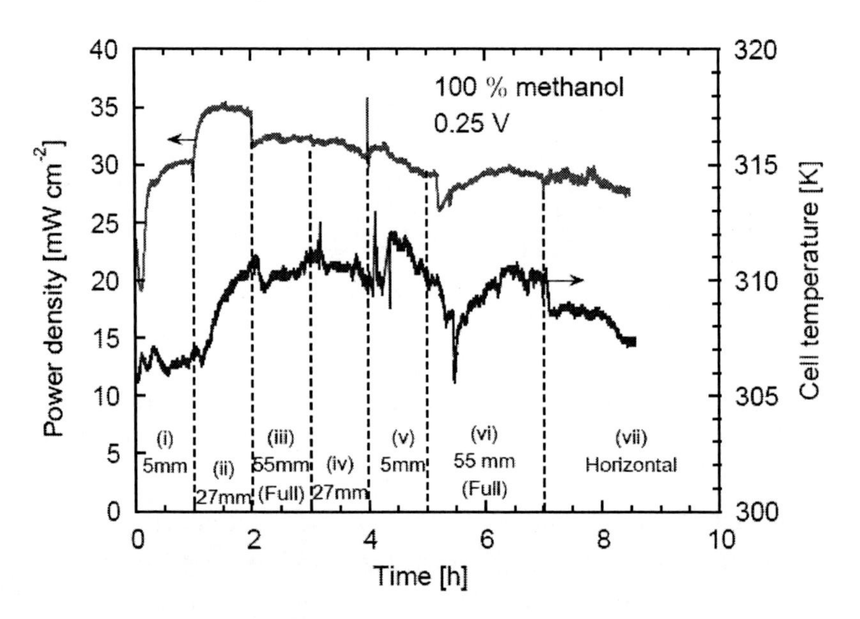

Figure 3-2. Influence of the methanol head height and cell orientation on the power generation and cell temperature with 100% methanol at the voltage of 0.25 V. Reproduced with permission from [21]. Copyright © 2012 Elsevier.

After the methanol reservoir was completely filled to 55 mm with the methanol solution in span (ii), the power density stabilized at 30 mW cm^{-2} regardless of the head height of the methanol although a slight variation in the cell temperature as seen after the methanol reservoir was completely filled to 55 mm with the methanol solution in span (iii). At 7 h (from span (vi) to span (vii)), we changed the cell orientation from vertical to horizontal while keeping the reservoir full. It was clear that the power density remained at 30 mWcm^{-2} for span (vi) and span (vii) as well as for spans (iii)-(vi). Therefore, it was confirmed that a nearly stable power density was obtained irrespective of the cell arrangement and the methanol head in the fuel reservoir in the vertically-oriented cell by the proper selection of the PCP properties.

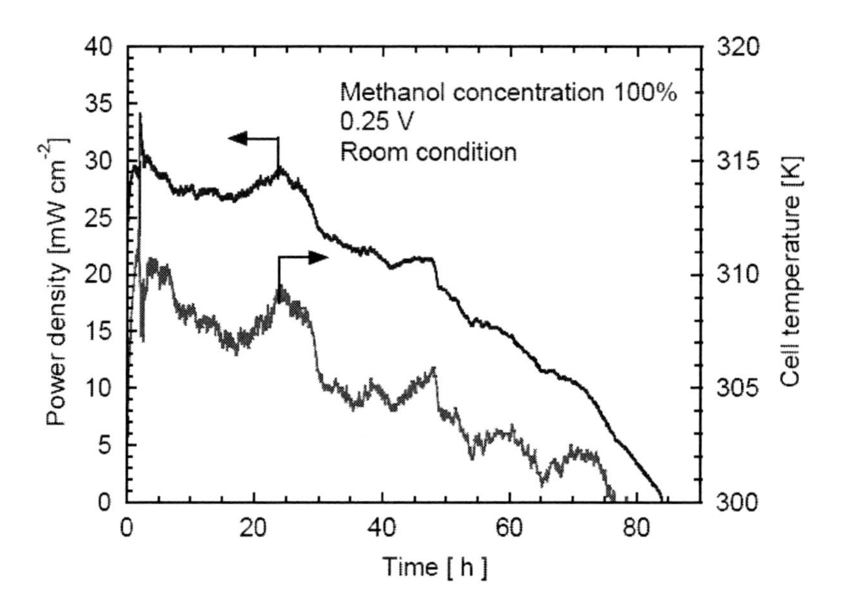

Figure 3-3. Power density and temperature profiles for long time operation with 100% methanol at a voltage of 0.25 V. Reproduced with permission from [21]. Copyright © 2012 Elsevier.

A long operation was also conducted using the 100% methanol. Figure 3-3 shows the long-term variations in the power density and the cell temperature measured at 0.25 V after injecting 12 cm^3 (55 mm in height) of 100% methanol. At 25 h, a power density of 28–30 mW cm^{-2} was obtained, then the power density gradually decreased with time. The decrease in the power density during this period might be due to the decreasing methanol concentration by the back diffusion of the water from the cathode to the anode [19]. Moreover, the decrease in the head height of the methanol in the reservoir might also cause the decrease in the power density after the 25-hour operation. Power generation was continued for 85 h until the methanol was completely consumed. During this operation, the volumetric energy density, based on the injected methanol volume (12 cm^3), was calculated to be about 800 Wh L^{-1}. This was higher than that obtained from other passive DMFCs of 600 Wh L^{-1}[16], 95 Wh L^{-1}[13] and 480 Wh L^{-1}[4]. In the vertically arranged single cell with a 55-mm height, an efficient operation was demonstrated. Hence, it is expected that a DMFC stack with a similar electrode structure using PCP could provide an efficient operation.

3.2. Passive cell stack with 8-single cells

Based on the successful results of the vertically-arranged single cell with the PCP, a passive DMFC stack was designed and fabricated as shown in Figure 3-4. It consisted of 8 unit cells, which had the same configuration of the single cell described in the previous section, and a common octagonal fuel reservoir for the eight cells with a total volume of 430 mL. The power generation characteristics were investigated under ambient conditions (303 K, 1 atm) using 100% methanol.

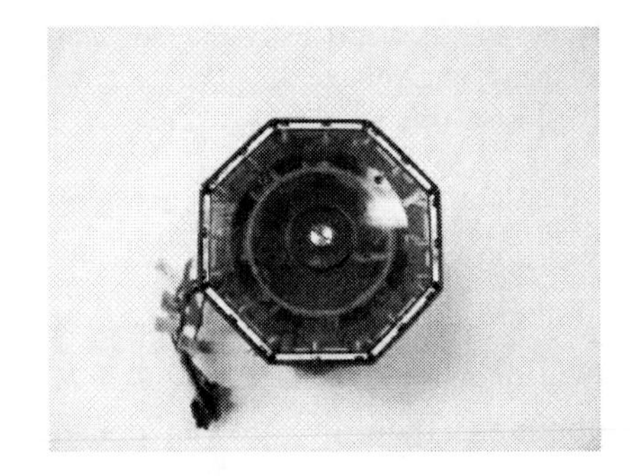

Figure 3-4. Pictures of the passive DMFC stack employing the fuel transport layer using the PCP. Reproduced with permission from [21]. Copyright © 2012 Elsevier.

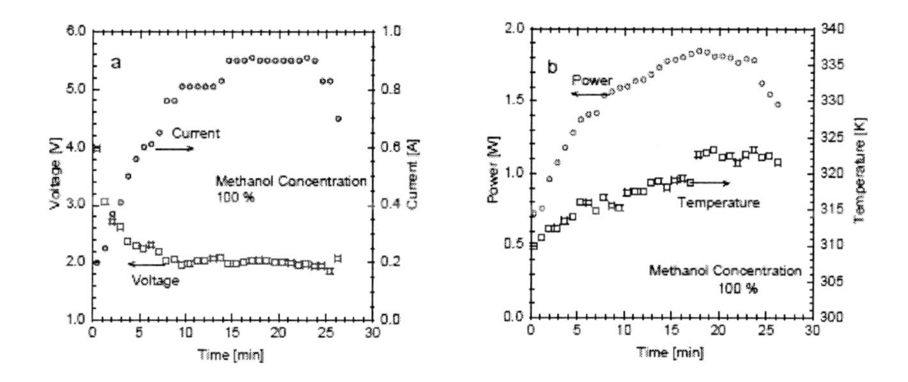

Figure 3-5. Power generation and temperature profiles of a passive DMFC stack with 100% methanol. (a) Current and voltage, (b) power and temperature. Reproduced with permission from [21]. Copyright © 2012 Elsevier.

Figure 3-5 shows the power generation profiles of the passive DMFC stack. In these experiments, the current was gradually increased to 0.9 A, then kept constant for a certain period as shown in Figure 3-5a. In Figure 3-5b, the power output increased with the increased current. When the current was set at 0.9 A, the power reached about 1.8 W. The maximum power output, 1.8 W, was calculated to be 27 mW cm^{-2}. The cell temperature increased with the increasing current as shown in Figure 3-5b. The cell temperature increased to 320 K, which was higher than that obtained from the single cell operations shown in Figure 3-2. This could be related to the low heat dissipation due to the compact stack design, i.e., high electrode area to volume, and/or the increase in the MCO of each unit cell in comparison to that in the single cell structure. Therefore, the polymer electrolyte membrane for the stack operation tended to be dehydrated at the higher temperature compared to that for the single cell operation, then the power density of the stack was lower than that of the single cell. Another reason for the lower power density, 27 mW cm^{-2}, of the stack compared to the single cell of 30 mW cm^{-2} would be explained by the uneven performance among the 8 unit cells as shown Figure 3-6. Figure 3-6 shows the cell voltages of the unit cells, cell 1 to cell 8, during the operation shown in Figure 3-5. One could see the difference in the cell-voltage among the 8 unit cells. The difference was relatively small, about

80 mV at 1 min, but gradually increased with time to about 0.125 V at 25 min. It is important to make each unit cell performance uniform during a DMFC stack operation [56-58] for a higher power output. The voltage distribution among the unit cells could be caused by many factors such as a small difference in the characteristics of the PCP and the hydration of the membrane. Although this DMFC stack was fabricated using the same type of PCP, there might be a small difference in the pore structure and/or pore distribution over the plate among the 8 unit cells. The difference would affect the methanol transport, then the unit cell performance and MCO amount. An increase in the stack temperature was observed during the stack operation as shown in Figure 3-5b. In order to achieve a higher power output for the DMFC stack, it is necessary to make sure that the performance of each unit cell is similar.

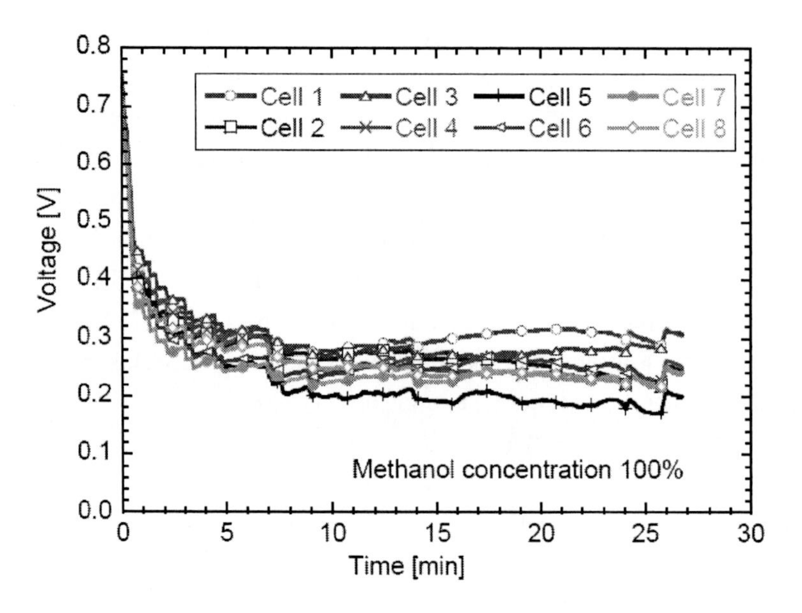

Figure 3-6. Comparison of the cell voltages of each single cell. Reproduced with permission from [21]. Copyright © 2012 Elsevier.

4. Development of an Active DMFC Stack with the PCP

4.1. Single Cell Performance of an Active DMFC with the PCP

4.1.1. Cell Structures and Controllers

Vapor-feed DMFCs that can use high methanol concentrations have been proposed as a passive or a semi-passive system [38, 59, 60]. There are currently no active DMFCs that operate with neat methanol. In the passive or semi-passive systems, the operating conditions including temperature, flow rate of the fuel and that of air as well as the air humidity are not consciously controlled, thus the power density of the passive DMFC was low. For the power generation system, an active system with control of the operating conditions is preferred for achieving a high-power density and high energy-conversion efficiency. Moreover, in a liquid feed DMFC system, water flooding usually occurs so that a high flow rate of air, about twenty times the stoichiometry, is sometimes fed to the cathode in order to remove the liquid water droplets and avoid cathode flooding. Under such conditions, almost one-third of the electric power generated by the DMFC is consumed for pumping air to the cathode side [61]. An alternative active DMFC system using the PCP is expected to provide an efficient power generation system with a reduced parasitic load for the air feed and liquid fuel supply at the cathode and anode, respectively. This could be related to the use of a high methanol concentration on the anode side and the absence of flooding on the cathode side by application of the fuel supply layer as already mentioned.

In these contexts, an active DMFC using the PCP with a 30 cm^2 active electrode area, was fabricated based on the studies of the passive systems.

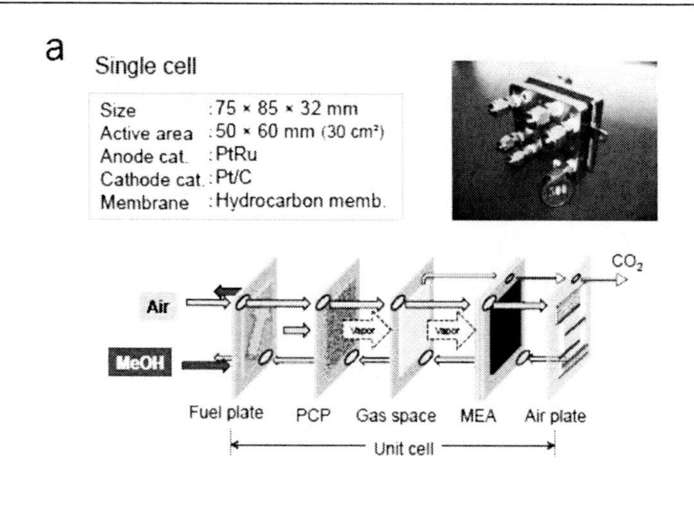

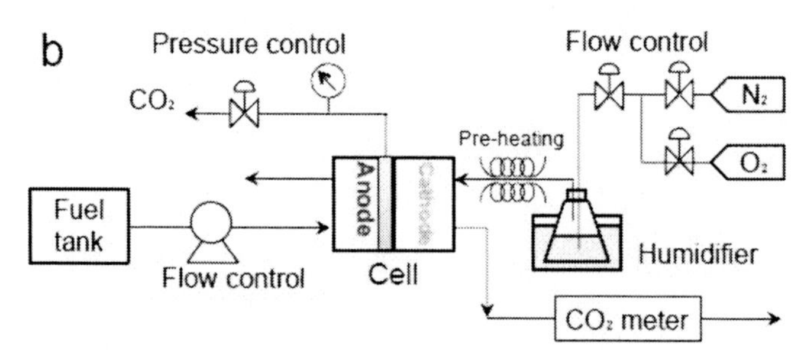

Figure 4-1. Schematic diagram of the active DMFC with the PCP (a) and typical controllers used for the operation (b). Reproduced with permission from [23]. Copyright © 2012 Elsevier.

Figure 4-1 shows a schematic diagram of the active DMFC with the PCP, (a), and typical controllers used for the operation, (b). The cell was designed and fabricated in a cooperative study with Chemix Co., Ltd., Japan. The cell consisted of an air plate, an MEA, a gas spacer, a porous carbon plate (PCP) and a fuel plate. The active area of the MEA was 30 cm^2 and was 6 times larger than that used in our previous test cells [19-21]. The PCP and the gas spacer were placed between the anode of the MEA and the fuel plate. All of the plates and the gas spacer were made of carbon. The air plate had flow channels of 2.0 mm width and 0.8 mm depth, and the fuel plate had a methanol flow-field. The single cell was designed as a stack unit. The thickness of the PCP and the spacer was 0.5

and 1.5 mm, respectively, then the total thickness of the unit, including the fuel plate and the air plate, was 6 mm. The PCP had small pores with the average size of a few micrometers. The sucking ability is not important for the PCP, in contrast to the case of the vertically-oriented passive DMFC, because the entire surface of the PCP is always in contact with the methanol solution fed through the flow channel. The cell temperature was controlled using rod heaters at the end plates, made of stainless steel, and running cooling water through them, which are not shown in the figure. In this cell structure, when the power generation started by feeding a methanol solution to the methanol flow-field, the CO_2 gas produced at the anode accumulated in the gap space and formed a gas layer between the anode and the PCP. A part of the gas in the gas layer can be discharged through a hole and a back-pressure valve as shown in Figure 4-1b.

4.1.2. Effect of the Discharging Gas Pressure and the Methanol Concentration

The pressure of the discharging gas from the anode gas space [22] is added to the operational parameters. The effect of the discharging gas pressure on the power generation characteristics is shown in Figure 4-2. The overall DMFC efficiency as the vertical axis was calculated based on the cell performance including the methanol crossover as follows:

$$\text{DMFC efficiency} = \left(\frac{\Delta G^{O}}{\Delta H^{O}}\right)\left(\frac{V_{cell}}{V_{the}}\right)\left(\frac{i_{cell}}{i_{cell}+i_{MCO}}\right) \quad (1)$$

where $\Delta G°$ is the standard Gibbs free-energy change for the methanol oxidation reaction:

$$CH_3OH_{(L)} + \frac{3}{2}O_2 \rightarrow CO_2 + 2H_2O_{(L)} \quad (2)$$

and $\Delta H°$ is the standard enthalpy change of the reaction, V_{cell} is the cell voltage, V_{the} is the theoretical cell voltage, and i is the current density of the fuel cell. The crossover current density, i_{MCO}, was calculated from the rate of the measured methanol crossover.

The pressure slightly affected the current–voltage (i–V) and the current–power density (i–P), Figure 4-2a. Although a relatively high power density of 42 mW cm^{-2} compared to that of the passive cell, Figure 3-2, was obtained at the medium pressure of 15 kPa (gauge pressure), the direct effect of the pressure on the kinetics of the electrode reactions would not be significant due to the small pressure change from 10 kPa to 30 kPa. On the other hand, the MCO decreased with increasing the pressure (Figure 4-2b), in accordance with that observed for the previous passive DMFC with PCP [22]. The pressure change would affect the position and area of the gas/water interface at the PCP. The higher pressure would increase the distance of methanol diffusion through the pores. Since the best power density and the best efficiency were obtained at a 15 kPa pressure, the pressure was set at 15 kPa for the operation.

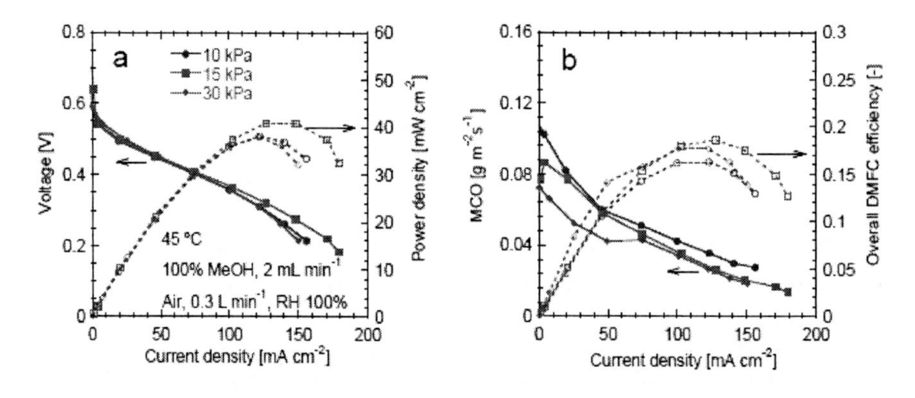

Figure 4-2. Performance of the active DMFC with the PCP obtained for the different discharging gas pressures, 10, 15 and 30 kPa, at 100% methanol (2 mL min^{-1}), air (RH 100%, 0.3 L min^{-1}), 45°C. Relationship between the current density and cell voltage as well as the power density, (a), and MCO with the overall DMFC efficiency, (b). Reproduced with permission from [23]. Copyright © 2012 Elsevier.

The DMFC showed the highest power density with a 100% (24.7 M) methanol as shown in Figure 4-3. The highest power density at 45°C was 40 mW cm^{-2} which was higher than that obtained by the vertically-oriented passive DMFC (Figure 3-2). This means that the structure of the fuel transport layer was successfully adjusted for the 100% methanol use under the active operating conditions. The figure for the DMFC efficiency,

Figure 4-3b, revealed that the operating condition for the highest power density was not that for the best efficiency. The MCO increased with the increasing methanol concentration, then the operation at 100% methanol produced the highest MCO resulting in the lowest efficiency. It was different compared to that of the passive DMFC with air breathing; the relationship between the i-V or i-P performance and the MCO or the efficiency was almost independent and not strongly related to each other. This would be due to the relatively high air flow rate, 0.3 L min^{-1}, that corresponds to 5 times the stoichiometry at 100 mA cm^{-2}. The air-flow rate eliminated the effect of methanol crossover at the cathode, although the air-flow rate of 5 times the stoichiometry was much lower than that used for the conventional liquid feed DMFCs with dilute methanol solutions.

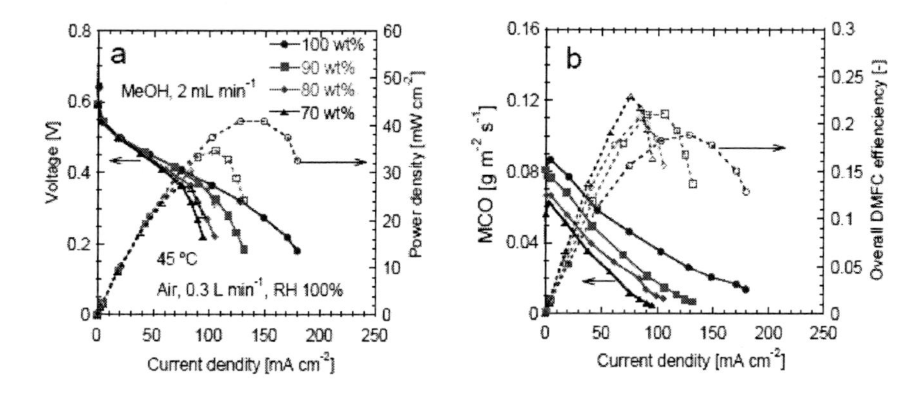

Figure 4-3. Effect of the methanol concentration on the active DMFC perfromance at 45°C and 100% methanol. Relationship between the current density and the cell voltages as well as the power desnsity, (a), and MCO with the overall DMFC efficiency, (b). Reproduced with permission from [23]. Copyright © 2012 Elsevier.

4.1.3. Power Generation Performance at 100% Methanol

The methanol flow rate slightly affected the performance (Figure 4-4), since the methanol flow along the outer surface of the PCP and perpendicular to the pores slightly affected the methanol transport through the PCP. The lowest flow rate of 2 mL min^{-2} in the figure (Figure 4-4) corresponded to 27 times the stoichiometry at 100 mA cm^{-2}. It is still too high and can be further reduced. We have confirmed that the DMFC can be operated without a fuel pump by just keeping the liquid head of the fuel in the tank higher than the top of the electrode. Theoretically, 100% methanol does not need a fuel circulation system because the methanol concentration of the anode, in principle, does not change under such conditions. Such a low flow rate is advantageous for the DMFC system to save the auxiliary power for the fuel supply and then increase the system efficiency.

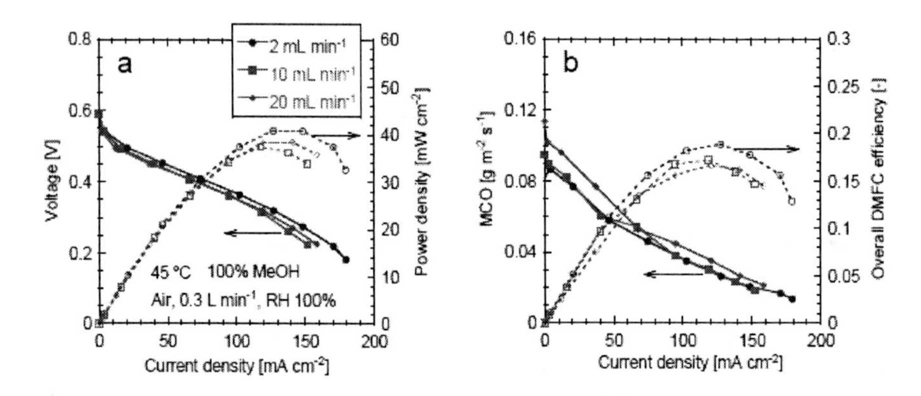

Figure 4-4. Effect of the methanol flow rates on the active DMFC perfromance with air (RH 100%, 0.3 L min^{-1}) at 45 °C. The methanol solution was 100% methanol. Relationship between the current density and cell voltages as well as the power desnsity, (a), and MCO with the overall DMFC efficiency, (b). Reproduced with permission from [23]. Copyright © 2012 Elsevier.

The air-flow rate affected the i-V performance especially at the higher current densities. The air-flow rate of 0.2 L min^{-1} corresponds to 3.5 times the stoichiometry at 100 mA cm^{-2}. The higher air-flow rate reduces the overvoltage of the oxygen transport to the cathode, and increases the limiting current density, thereby the higher power density (Figure 4-5). However, the higher air-flow rate causes the higher methanol crossover thus the lower efficiency. The increase in the air-flow rate promotes removal of the methanol and water from the cathode and causes an increase in the MCO. The similar i-V curves at the low current densities below 50 mA cm^{-2} for the different air-flow rates despite the big difference in the MCO also suggested the removal of the product water from the cathode by the air-flow that was more than 3 times greater than that of the stoichiometry at 100 mA cm^{-2}. For comparison, the liquid feed DMFC that is usually operated with a high air-flow rate, sometimes more than 20 times that of the stoichiometry [61], while the DMFC with the PCP can be operated at a lower air-flow rate. The lower air-flow rate saves the auxiliary power necessary for the air supply in the DMFC system, therefore, increases the system efficiency.

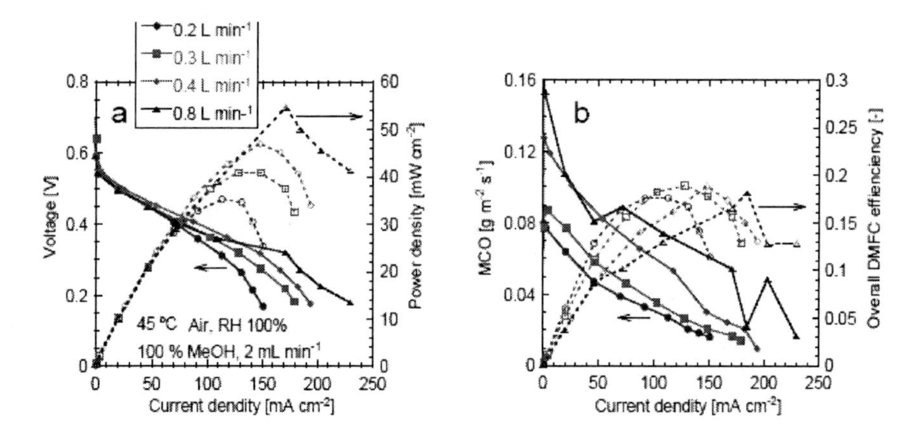

Figure 4-5. Effect of the air-flow rate on the active DMFC performance with 100% methanol (2 mL min^{-2}) at 45 °C. Relationship between the current density, the cell voltage as well as the power density, (a), and MCO with the overall DMFC efficiency, (b). Reproduced with permission from [23]. Copyright © 2012 Elsevier.

The air humidity slightly affected the i-V performance in the humidity range from 30% to 100%. However, it significantly affected the MCO and the DMFC efficiency (Figure 4-6). The MCO decreased with the increasing humidity. For the DMFC with the PCP, the water transport occurs from the cathode to the anode due to the deficiency of water for the electrode reaction at the anode. The higher water flux to the anode at the higher humidity would reduce the MCO.

Generally, the liquid feed DMFC operated at the low concentration methanol is slightly influenced by the air humidity because of the highly-hydrated membrane of this system. The strong effect of the air humidity on the MCO is a characteristic of the DMFC with the PCP. It is necessary for the DMFC with the PCP to keep the air humidity high, such as RH 100%, for the efficient power generation. Keeping the air humidity at such a high

RH is easily achieved in the DMFC with the PCP because the total reaction of the methanol oxidation occurring in the DMFC is water production.

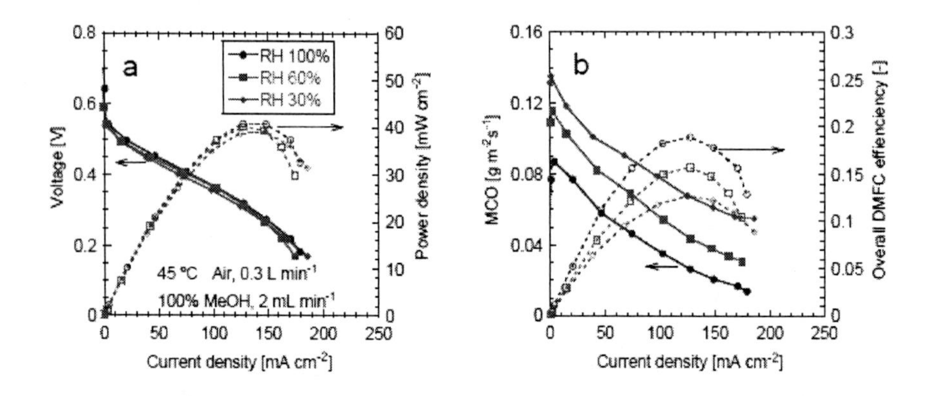

Figure 4-6. Effect of the air humidity on the active DMFC performance with 100% methanol (2 mL min^{-1}) at 45 °C. The air-flow rate was 0.3 L min^{-1}. Relationship between the current density and cell voltage as well as the power density, (a), and MCO with the overall DMFC efficiency, (b). Reproduced with permission from [23]. Copyright © 2012 Elsevier.

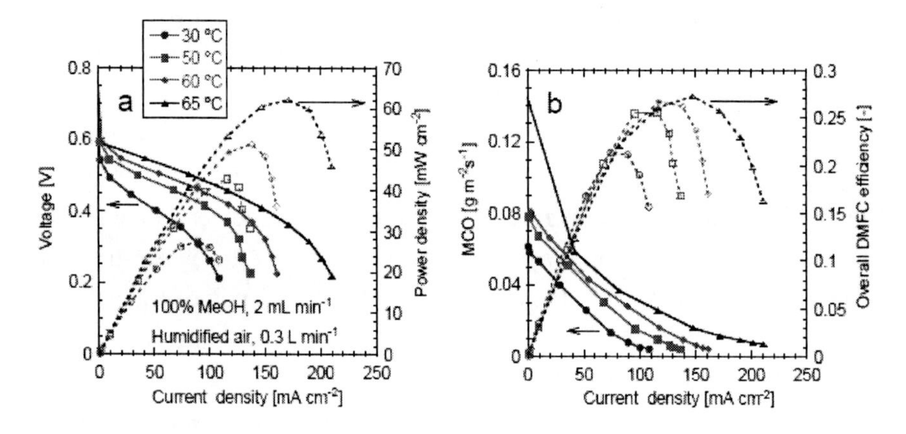

Figure 4-7. Effect of the temperature on the active DMFC performance with 100% methanol (2 mL min^{-1}) and humidified air (0.3 L min^{-1}). Relationship between the current density and cell voltage as well as the power density, (a), and MCO with the overall DMFC efficiency, (b). Reproduced with permission from [23]. Copyright © 2012 Elsevier.

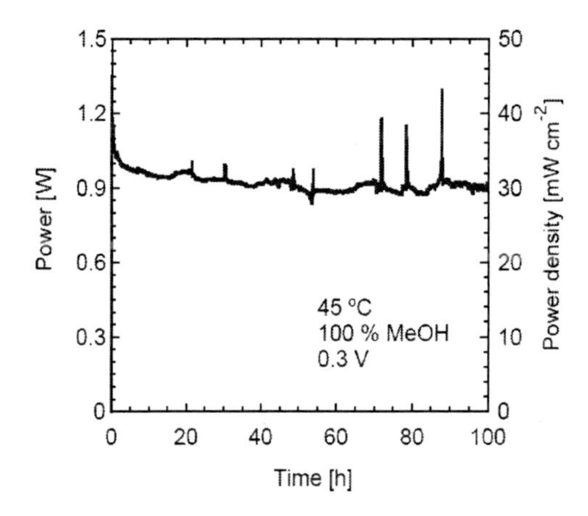

Figure 4-8. Power and power density with time during the constant voltage operation using 100% methanol at 0.3 V and 45°C for 100 h. Reproduced with permission from [23]. Copyright © 2012 Elsevier.

The effect of the operating temperature on the performance of the DMFC is shown in Figure 4-7. The temperature had a significant effect on the power density. The higher the temperature, the higher the i–V performance and higher the power density. These were explained by the increase in the electrode reactions rates for both the anode and cathode and the increased ionic conductivity of the electrolyte with the increase in the cell temperature. The DMFC efficiency increased with the increase in the cell temperature and reached 62 mW cm^{-2} at 65°C, that was more than double that at 30°C. On the other hand, the MCO also increased with the increase in the temperature due to the increased methanol permeability with the increase in temperature. The effect of the increase in the MCO was not related to the power density.

Figure 4-8 shows the power density profile at the constant cell voltage of 0.3 V and 45°C for 100 h. During the long-term operation, although the initial power density of about 40 mW cm^{-2} decreased to 30 mW cm^{-2} within 5 h, the power density remained almost constant for 100 h. The several noise spikes that appeared in the figure was related to the methanol refilling. This figure confirmed that the DMFC with the PCP could be stably operated for a long time. Since the noise spikes that appeared showed an increased power density, the degradation may be related to limiting the supply of the methanol or oxygen to the electrodes.

4.2. Development of an active DMFC Stack with the PCP and Demonstration

We assembled a short stack consisting of 5 single cells and evaluated its power generation performance. Figure 4-9 shows images of the 5-cell stack. The single cell component of the stack was exactly that of the single cell described in the last section. The dimensions of the 5-cell stack were 75 mm in height, 85 mm in width and 55 mm in length.

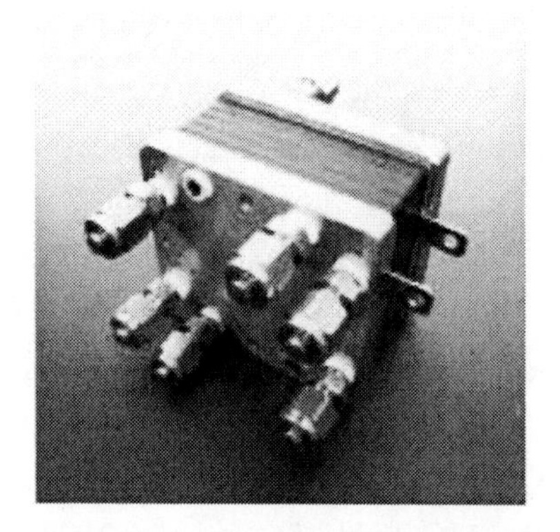

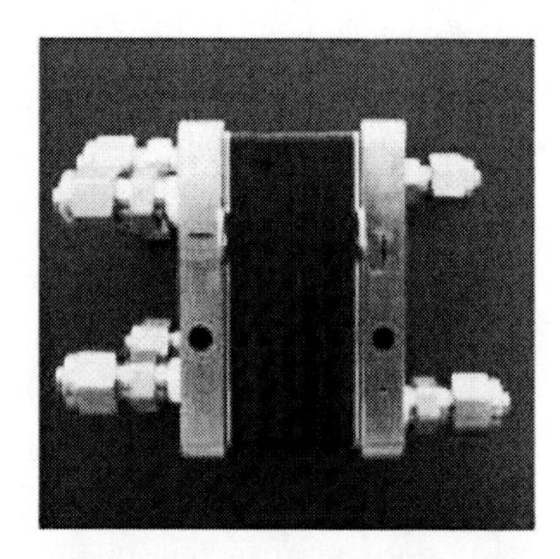

Figure 4-9. Image of the 5-cell short stack employing the fuel transport layer using the PCP.

Figure 4-10a shows the power generation performance of the 5-cell stack operated with 100% methanol for over 24 hours. The operating conditions are shown in Table 4-1. At the start, 100 ml of 100% methanol was poured to the fuel tank and another 100 ml of the methanol was added to the fuel tank after 14 h of operation since the level of the methanol in the tank decreased. A 5 W (33.3 mW cm^{-2}) power output that corresponded to that of the single cell was initially obtained. The power output was quite stable during the operation. The stable operation was achieved based on the uniform performance among the 5 unit-cells shown in Figure 4-10b as a result of the uniformly controlled temperatures. The slight decrease in the power density during the 24-h operation was related to the decreased MCO.

Table 4-1. Operating conditions of the 5-cell stack

Anode	100% methanol, 100 ml (Initial), 10 ml min^{-1}
Cathode	Air 1.2 L min^{-1}, RH 100%
Cooling water	48°C, 80 mL min^{-1}
Operation mode	Constant current, 2.5 A

The decrease of the MCO with time would be related to the decrease of the methanol concentration in the flow channels. There was a water flux from the anode to the methanol flow channel based on the concentration difference between them, although the flux was low. During the DMFC operation, the methanol concentration was reduced by the water flux. Actually, the measured concentration of the fuel in the tank slightly deceased with time, i.e., 97wt% and 96wt% at 6 h and 14 h, respectively. The local water concentration at the gas/water interface of the PCP, where the methanol was vaporized, is higher than that in the fuel tank. Therefore, the decrease in the methanol concentration reduced the methanol flux through the PCP which then reduced the power output.

The employment of a water-resisting film in the fuel transport layer would be effective to resist the reduction of the methanol concentration and keep the power density more stable, but this was not investigated. In a recent report, F. Zhang et al. reported that the water-resisting film, a pervaporation film, applied to the fuel supply layer, restricted the water transport and stabilized the DMFC power density for 400 h [62]. Hence, we can expect further improvement of the stability of the DMFC power output by adding the water-resisting film to the fuel transport layer.

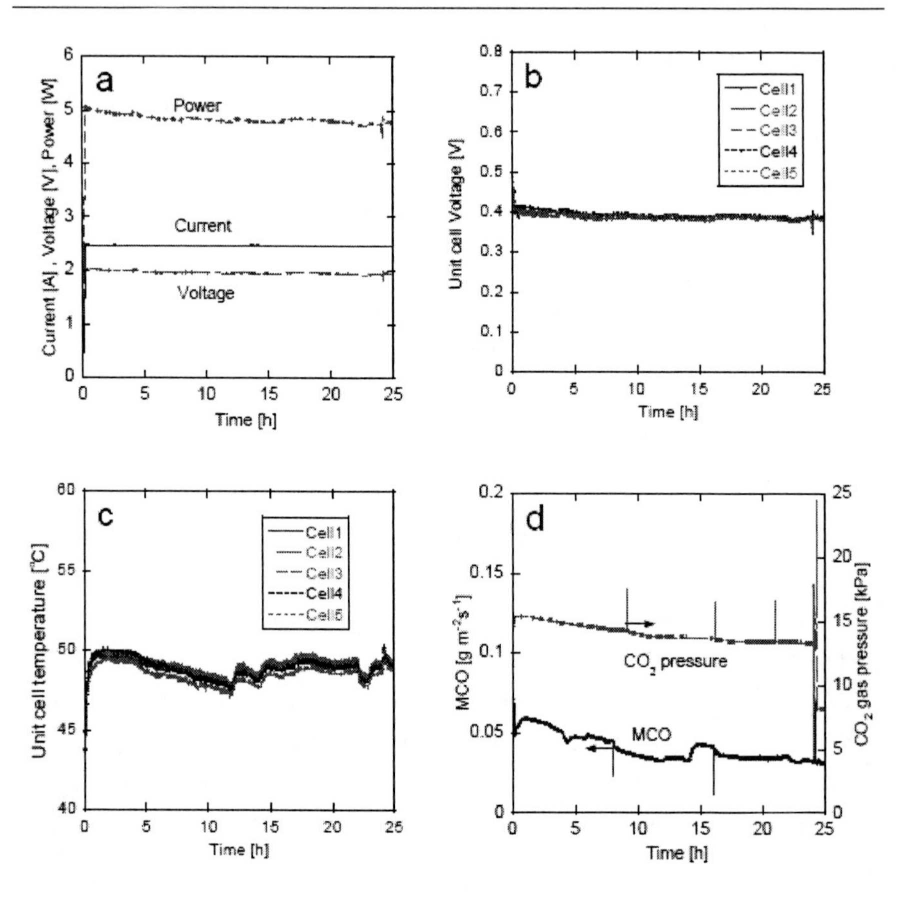

Figure 4-10. The results of the 5-cell stack operation for 24 h. (a) Power, current and voltage, (b) Unit cell voltage, (c) Unit cell temperature, (d) CO_2 pressure and MCO.

CONCLUSION

In this chapter, we introduced our study of the fuel supply layer using PCP for the utilization of highly concentrated methanol up to neat methanol in a DMFC. The controlled mass transport based on the various PCP properties and the layer dimensions, the single cell performance of the horizontally-oriented passive DMFC and an extension to the vertical cell orientation were also reported. Furthermore, both a passive DMFC stack with the PCP and an active DMFC stack with the PCP were demonstrated.

The technique using the fuel supply layer with the PCP is quite effective for increasing the energy density of a DMFC power generation system.

The power density of our DMFC with the PCP was at most 40–60 mW cm^{-2} and may not be high enough for practical applications. However, a high power density can be expected by combining the fuel transport layer with a high performance MEA. Wan recently demonstrated a very high DMFC power generation, over 300 mW cm^{-2}, using a thin electrolyte membrane [63]. The methanol barrier for the MEA with a high performance can be optimized by an appropriate pore structure of the PCP and the layer structure. Such a combination provides a DMFC power generation system; it not only increases the energy density, but also increases the power density.

The methanol and water flux through the MEA significantly differed by employing the fuel transport layer that can be reproduced by calculations using the proper numerical model. A. Fahrig et al. developed a two-dimensional, two-phase and non-isothermal model for a passive vapor-feed DMFC with a membrane vaporizer [64]. Such a numerical model would be effective in designing the dimensional parameters of the fuel transport layer and to estimate the power generation performance for specific operating parameters.

The fuel transport layer with the PCP is effective for controlling the mass transport through the MEA, and the DMFC with the PCP make highly concentrated methanol, up to neat methanol, use possible, and it could be applied not only for a passive system, but also for an active system.

REFERENCES

[1] Badwal, S.P.S., Giddey, S.S., Munnings, C., Bhatt, A., Hollenkamp, A.F., (2014). Emerging electrochemical energy conversion and storage technologies. *Front. Chem.* 2, 1-28.

[2] Arico, A.S., Srinivasan, S., Antonucci, V., (2001). DMFCs: From Fundamental Aspects to Technology Development. *Fuel Cells* 1, 133–161.

[3] Schultz, T., Zhou, S., Sundmacher, K., (2001). Current Status of and Recent Developments in the Direct Methanol Fuel Cell. *Chem. Eng. Technol.* 24, 1223–1233.

[4] Liu, J.G., Zhao, T.S., Chen, R., Wong, C.W., (2005). The effect of methanol concentration on the performance of a passive DMFC. *Electrochem. Commun.* 7, 288-294.

[5] Liu, J.G., Zhao, T.S., Liang, Z.X., Chen, R., (2006). Effect of membrane thickness on the performance and efficiency of passive direct methanol fuel cells. *J. Power Sources* 153, 61–67.

[6] Wang, J.T., Wasmus, S., Savinell, R.F., (1996). Real-Time Mass Spectrometric Study of the Methanol Crossover in a Direct Methanol Fuel Cell. *J. Electrochem. Soc.* 143, 1233–1239.

[7] Peled, E., Duvdevani, T., Aharon, A., Melman, A., (2000). A Direct Methanol Fuel Cell Based on a Novel Low-Cost Nanoporous Proton-Conducting Membrane. *Electrochem. Solid St.* 3, 525–528.

[8] Fedkin, M.V., Zhou, X., Hofmann, M.A., Chalkova, E., Weston, J.A., Allcock, H.R., Lvov, S.N., (2002). Evaluation of methanol crossover in proton-conducting polyphosphazene membranes. *Mater. Lett.* 52, 192–196.

[9] Yamaguchi, T., Ibe, M., Nair, B.N., Nakao, S., (2002). A Pore-Filling Electrolyte Membrane-Electrode Integrated System for a Direct Methanol Fuel Cell Application. *J. Electrochem. Soc.* 149, A1448–A1453.

[10] Ponce, M.L., Prado, L., Ruffmann, B., Richau, K., Mohr, R., Nunes, S.P., (2003). Reduction of methanol permeability in polyetherketone–heteropolyacid membranes. *J. Membr. Sci.* 217, 5–15.

[11] Lu, G.Q., Wang, C.Y., Yen, T.J., Zhang, X., (2004). Development and characterization of a silicon-based micro direct methanol fuel cell. *Electrochim. Acta* 49, 821-828.

[12] Guo, Z., Faghri, A., (2006). Miniature DMFCs with passive thermal-fluids management system. *J. Power Sources* 160, 1142-1155.

[13] Kim, H., (2006). Passive direct methanol fuel cells fed with methanol vapor. *J. Power Sources* 162, 1232–1235.

[14] Nakagawa, N., Kamata, K., Nakazawa, A., Abdelkareem, M.A., Sekimoto, K., (2006). Methanol crossover controlled by a porous carbon plate as a support. *Electrochemistry* 74, 221-225.

[15] Nakagawa, N., Abdelkareem, M.A., Sekimoto, K., (2006). Control of methanol transport and separation in a DMFC with a porous support. *J. Power Sources* 160, 105-115.

[16] Abdelkareem, M.A., Nakagawa, N., (2006). DMFC employing a porous plate for an efficient operation at high methanol concentrations. *J. Power Sources* 162, 114–123.

[17] Abdelkareem, M.A., Nakagawa, N., (2007). Effect of oxygen and methanol supply modes on the performance of a DMFC employing a porous plate. *J. Power Sources* 165, 685–691.

[18] Abdelkareem, M.A., Morohashi, N., Nakagawa, N., (2007). Factors affecting methanol transport in a passive DMFC employing a porous carbon plate. *J. Power Sources* 172, 659–665.

[19] Nakagawa, N., Abdelkareem, M.A., (2007). The Role of Carbon Dioxide Layer Prepared by a Porous Carbon Plate in a Passive DMFC as a Mass Transport Barrier. *J. Chem. Eng. Jpn* 40, 1199-1204.

[20] Abdelkareem, M.A., Yoshitoshi, T., Tsujiguchi, T., Nakagawa, N., (2010). Vertical operation of passive direct methanol fuel cell employing a porous carbon plate. *J. Power Sources* 195, 1821–1828.

[21] Tsujiguchi, T, Abdelkareem, A.M., Kudo, T., Nakagawa, N., (2010). Development of a passive direct methanol fuel cell stack for high methanol concentration. *J. Power Sources* 195, 5975–5979.

[22] Masdar, M.S., Tsujiguchi, T., Nakagawa, N., (2010). Improvement of water management in a vapor feed direct methanol fuel cell. *J. Power Sources* 195, 8028–8035.

[23] Nakagawa, N., Tsujiguchi, T., Sakurai, S., Aoki, R., (2012). Performance of an active direct methanol fuel cell fed with neat methanol. *J. Power Sources* 219, 325-332.

[24] Surampudi, S., Narayanan, S.R., Vamos, E., Frank, H., Halpert, G., (1994). Advances in direct oxidation methanol fuel cells. *J. Power Sources* 47, 377-385.

[25] Ravilumar, M.K., Shukla, A.K., (1996). Effect of Methanol Crossover in a Liquid-Feed Polymer-Electrolyte Direct Methanol Fuel Cell. *J. Electrochem. Soc.* 143, 2601-2606.

[26] Awang, N., Ismail, A.F., Jaafar, J., Matsuura, T., Junoh, H., Othman, M.H.D, Rahman, M.A., (2015). Functionalization of polymeric materials as a high performance membrane for direct methanol fuel cell: A review. *React. Funct. Polym.* 86, 248-258.

[27] Siroma, Z., Fujiwara, N., Ioroi, T., Yamazaki, S, Yasuda, K., Miyazaki, Y., (2004). Dissolution of Nafion® membrane and recast Nafion® film in mixtures of methanol and water. *J. Power Sources* 126, 41-45.

[28] Mallick, R. K., Thombre, S. B., Shrivastava, N. K., (2016). Vapor feed direct methanol fuel cells (DMFCs): A review. *Renew. Sust. Energy Rev.* 56, 51-74.

[29] Shukla, A.K., Christensen, P.A., Hamnett, A., Horgarth, M.P., (1995). A vapor feed direct methanol fuel cell with proton exchange membrane electrolyte. *J. Power Sources* 55, 87–91.

[30] Scott, K., Taama, W.M., Argyropoulos, P., (1999). Engineering aspects of the direct methanol fuel cell system. *J. Power Sources* 79,43–59.

[31] Kallo, J., Wernes, L., Helmolt, R.V., (2003). Conductance and methanol crossover investigation of Nafion membrane in a vapor feed DMFC. *J. Electrochem. Soc.* 150, A765– A769.

[32] Guo, Z., Faghri, A., (2007). Vapor feed direct methanol fuel cells with passive thermal fluids management system. *J. Power Sources* 167, 378–90.

[33] Wang, J.T., Wainright, J.S., Savinell, R.F., Litt, M., (1996) A direct methanol fuel cell using acid doped polybenzimidazole as polymer electrolyte. *J. Appl. Electrochem.* 26, 751–756.

[34] Tsujiguchi, T., Masumi, N., Kawakubo, A., Nakagawa, N., (2010). Control of Methanol Crossover Using a Perforated Metal Sheet for DMFC Application. *Key Eng. Mater.* 459, 71-77.

[35] Pan, Y.H., (2006). Direct methanol fuel cell with concentrated solutions. *Electrochem. Solid St.* 9, A349–A351.

[36] Xu, C., Faghri, A., Li, X., (2011). Improving the water management and cell performance for the passive vapor-feed DMFC fed with neat methanol. *Int. J. Hydrogen Energy* 36, 8468-8477.

[37] He, Y.L., Miao, Z., Yang, W.W., (2012). Characteristics of heat and mass transport in a passive direct methanol fuel cell operated with concentrated methanol. *J. Power Sources* 208, 180–186.

[38] Eccarius, S., Krause, F., Beard, K., Agert, C., (2008). Passively operated vapor-fed direct methanol fuel cells for portable applications. *J. Power Sources* 182, 565–579.

[39] Halim, F.A., Hasram, U.A., Masdar, M.S, Kamarudin, S.K., (2013). Study on a passive vapor feed direct methanol fuel cell with high methanol concentration. *J. Clean Technol.* 1, 292–294.

[40] Guo, J., Zhang, H., Jiang, J., Huang, Q., Yuan, Tang, H., (2013). Development of a self adaptive direct methanol fuel cell fed with 20M methanol. *Fuel Cells* 6, 1018–1023.

[41] Yuan, W., Tang, Y., Yang, X., Wan, Z., (2012). Toward using porous metal-fiber sintered plate as anodic methanol barrier in a passive direct methanol fuel cell. *Int. J. Hydrogen Energy* 37, 13510–13521.

[42] Wu, H., Zhang, H., Chen, P., Guo, J., Yuan, T., Zheng, J., (2014). Integrated anode structure for passive direct methanol fuel cells with neat methanol operation. *J. Power Sources* 248, 1264–1269.

[43] Peled, E., Blum, A., Aharon, A., Philosoph, M., Lavi, Y., (2003). Novel Approach to Recycling Water and Reducing Water Loss in DMFCs. *Electrochem. Solid St.* 6, A268-A271.

[44] Wu, Q.X., Zhao, T.S., Chen, R., Yang, W. W., (2010). Enhancement of water retention in the membrane electrode assembly for direct methanol fuel cells operating with neat methanol. *Int. J. Hydrogen Energy* 35, 10547-10555.

[45] Wu, Q.X., Zhao, T.S., (2011). Characteristics of water transport through the membrane in direct methanol fuel cells operating with neat methanol. *Int. J. Hydrogen Energy* 36, 5644-5654.

[46] Zhang, Z., Yuan, W., Wang, A., Yan, Z., Tang, Y., Tang, K., (2015). Moisturized anode and water management in a passive vapor feed direct methanol fuel cell operated with neat methanol. *J. Power Sources* 297, 33–44.

[47] Xiu, Y., Kamata, K., Nakagawa, N., (2002). Gas evolution at the anode and the coulombic efficiency of DMFC. *Electrochemistry* 70, 946-949.

[48] Sanicharane, S., Bo, A., Sompalli, B., Gurau, B., Smorkin, E.S., (2002). *In Situ* 50°C FTIR Spectroscopy of Pt and PtRu Direct Methanol Fuel Cell Membrane Electrode Assembly Anodes. *J. Electrochem. Soc.* 149, A554-A557.

[49] Liua, R., Fedkiw, P.S., (1992). Partial Oxidation of Methanol on a Metallized Nafion Polymer Electrolyte Membrane. *J. Electrochem. Soc.* 139, 3514-3523.

[50] Nakagawa, N., Sekimoto, K., Masdar, M.S., Noda, R., (2009). Reaction analysis of a direct methanol fuel cell employing a porous carbon plate operated at high methanol concentrations. *J. Power Sources* 186, 45-51.

[51] Masdar, M.S., Tsujiguchi, T., Nakagawa, N., (2009). Mass spectroscopy for the anode gas layer in a semi-passive DMFC using porous carbon plate Part I: Relationship between the gas composition and the current density. *J. Power Sources* 194, 610-617.

[52] Masdar, M.S., Tsujiguchi, T., Nakagawa, N., (2009). Mass spectroscopy for the anode gas layer in a semi-passive DMFC using porous carbon plate Part II: Relationship between the reaction products and the methanol and water vapor pressures. *J. Power Sources* 194, 618-624.

[53] Chen, R., Zhao, T.S., Liu, J.G., (2006). Effect of cell orientation on the performance of passive direct methanol fuel cells. *J. Power Sources* 157, 351–357.

[54] Zhao, T.S. et al., (2009). Small direct methanol fuel cells with passive supply of reactants. *J. Power Sources* 191, 185–202.

[55] Lai, Q.Z., Yin, G.-P., Zhang, J., Wang, Z.B., Cai, K.D., Liu, P., (2008). Influence of cathode oxygen transport on the discharging time of passive DMFC. *J. Power Sources* 175, 458–463.

[56] Liu, Y., Xie, X., Shang, Y., Li, R., Qi, L., Guo, J., Mathur, V.K., (2007). Power characteristics and fluid transfer in 40 W direct methanol fuel cell stack. *J. Power Sources* 164, 322–327.

[57] Lee, S., Kim, D., Lee, J., Chung, S.T., Ha, H.Y., (2005). Comparative studies of a single cell and a stack of direct methanol fuel cells. *Korean J. Chem. Eng.* 22, 406–411.

[58] Kim, D., Lee, J., Lim, T.H., Oh, I.H., Yong, H., (2006). Operational characteristics of a 50W DMFC stack. *J. Power Sources* 155, 203–212.

[59] Li, X., Faghri, A., Xu, C., (2010). Water management of the DMFC passively fed with a high-concentration methanol solution. *Int. J. Hydrogen Energy* 35, 8690–8698.

[60] Xu, C., Faghri, A., Li, X., (2010). Development of a High Performance Passive Vapor-Feed DMFC Fed with Neat Methanol. *J. Electrochem. Soc.* 157, B1109–B1117.

[61] Hatanaka, T., (2002). Direct methanol fuel cell. *R&D Rev. Toyota CRDL* 37, 59-64 (Japanese).

[62] Zhang, F., Jiang, J., Zhou, Y., Xu, J., Huang, Q., Zou, Z., Fang, J., Yang, H., (2017). A Neat Methanol Fed Passive DMFC with a New Anode Structure. *Fuel Cells* 17, 315-320.

[63] Wan, N., (2017). High performance direct methanol fuel cell with thin electrolyte membrane. *J. Power Sources* 354, 167-171.

[64] Xu, C., Faghri, A., (2010). Mass transport analysis of a passive vapor-feed direct methanol fuel cell. *J. Power Sources* 195 7011-7024.

In: Direct Methanol Fuel Cells
Editors: Rose Hernandez et al.
ISBN: 978-1-53612-603-7

Chapter 3

TWO-PHASE FLOW PHENOMENA IN μDMFCS: A REVIEW OF MODELING AND VISUALIZATION TECHNIQUES

D. S. Falcão*, J. P. Pereira and A. M. F. R. Pinto*

CEFT, Departamento de Engenharia Química, Faculdade de Engenharia da Universidade do Porto, Porto, Portugal

ABSTRACT

Two-phase flow phenomena occur in low temperature direct methanol fuel cells through the generation of CO_2 gaseous bubbles at the anode side, or due to the accumulation of water droplets at the cathode side. In order to achieve better cell performances, two-phase flow characterization is crucial when designing these devices. Therefore, flow visualization is fundamental to better understand the effect of two-phase flows at macro- and especially at micro-scales. Particle Image Velocimetry (PIV) is a measurement technique that can be applied to both scales. This technique has received increased attention over the last two decades, since it allows the experimental validation of the multiphase behavior simulations usually performed by Computational Fluid Dynamics (CFD). In this chapter, a survey of the modeling approaches

* Corresponding Authors Email: dfalcao@fe.up.pt, apinto@fe.up.pt.

concerning multiphase flow in fuel cells is carried out. An attempt is made to review the flow visualization techniques applied for flow analysis in fuel cells, and a short description of the PIV methodology is provided, considering both macro and micro-scales. Additionally, the applications of this technique for the visualization and characterization of two-phase flow in fuel cells are dealt in detail. The advances made in this area are considered of increasing interest, since this technique may provide insight for the improved design and performance of fuel cells at the micro-scale.

Keywords: μDMFCs, μPIV, two-phase flow, visualization studies

INTRODUCTION

Fuel cells are a key component for sustainable energy, as they enable clean and efficient production of electricity from assorted primary sources. In the face of ever-increasing functionality and performance of portable consumer electronics, the integration of fuel cells into portable devices as revolutionary batteries is expected to occur in the next generation of mobile phones, PDAs, laptop computers and MP3 players. However, for widespread use in portable applications, the actual fuel cell systems need optimization to improve their cost, efficiency and durability [1, 2].

The use of micro-channels was proposed to have benefits for fuel cells such as improved convective transport, flexibility of design and decreased resistive path lengths. Micro- and nanofluidics is a rapid emergent field, and the research on this issue has been exponentially increasing over the last two decades. The micro-manufacturing processes coupled to the exploitation of micro-scale transport phenomena allowed the miniaturization of conventional fuel cells, leading to significantly cheaper and smaller, nevertheless high power density fuel cells [1, 2]. Currently, the micro-direct methanol fuel cells (μDMFCs) are the most promising power sources for portable electronics, especially due to the easy handling for fuel delivery, low operation temperature, noncorrosive electrolyte, and encouraging power densities that can reach up to ten times more than the conventional lithium-ion batteries [2-4]. Micro-direct methanol fuel cells have a simple design, centered on a membrane between bipolar plates where the flow channels are placed to supply fuel to the anode and oxidant to the cathode, respectively. On each side of the membrane are placed the catalyst layers (CLs), followed by the gas diffusion layers (GDLs), usually made of carbon fibers that create a porous diffusion medium. Although these devices are currently being exploited for numerous applications, the actual μDMFCs still have low efficiencies in the range of 20–25% [2], and thus several researchers have focused on the design parameters to optimize the flow distribution and enhance the cell performance [3-7]. Uniform flow distribution is highly advantageous to improve the performance of μDMFCs, since it has been demonstrated that the complex heat and mass transfer processes occurring within fuel cells are strongly dependent on the dynamics of fuel and oxygen [8-10]. At the micro-scale, the surface tension forces become dominant, especially considering liquid-vapor flows, and significantly affect the reactant dynamics when compared to the viscous, inertial and pressure forces [11].

For an adequate design of fuel cell systems, the prediction of two-phase flow is crucial. Opposed to the single-phase, multiphase systems deal with complex physicochemical phenomena, such as hydrodynamics of fluids with free surfaces, surface phenomena, besides the heat and mass transport coupled to the electrochemical reactions. Moreover, in two-phase flows the surface tension at the air-water interface severely influences the behavior of the fluid phases. Two-phase flows occur in low temperature direct methanol fuel cells through the generation of CO_2 gaseous bubbles at the anode side, or due to the accumulation of water droplets at the cathode side. The GDLs placed between the flow channels and the membrane make the two-phase flow phenomenon even more complex, as the CO_2 bubbles may block the cross-section of micro-channels, tending to form slugs that can move along the channels. Additionally, the GDLs are usually treated with Teflon (PTFE) solution, which increases the hydrophobicity of the medium and avoid the water droplets from reentering the GDL by agglomerating them [12]. This leads to the active catalyst sites obstruction, causing oxidant starvation and reducing the reactants mass transport velocity. Consequently, these occurrences affect the general cell performance, accelerate the cell degradation, and may lead to a dramatic decrease of the cell efficiency [4-13]. Several researchers have developed transport models to describe the two-phase flow phenomena, namely computational fluid dynamic (CFD) models considering slug flow in micro-channels [14-16], proton exchange membrane fuel cells (PEMFCs) [17-20], namely DMFCs [21-23] and, more recently, intensive efforts have focused on the modeling of micro-fuel cells [3, 4, 24-32]. However, the understanding of the two-phase flow is still reduced, especially in porous media such as the GDLs, and there is a need of comprehensive reviews of fluid mechanics in fuel cells. The present work provides a review of the important modeling studies regarding two-phase flow phenomena in the channels of DMFCs and μDMFCs. Furthermore, an overview of the most relevant flow visualization techniques employed for two-phase flow analysis in fuel cells is also presented, focusing on the PIV methodology considering both macro and micro-scales. Although other reviews exist concerning visualization at micro-scales [33-36], and regarding the μPIV methodology [37, 38], as far as the authors are concerned there is a lack of published information focusing on the applications of this technique in DMFCs and μDMFCs. The central purpose of this work is to gather and summarize the existing relevant two-phase flow modeling approaches and visualization experimental techniques, and thus to provide a useful guideline for further studies aiming the improved design of fuel cells at micro-scale.

MODELING STUDIES

The formulation of adequate mathematical models is based on the need for a deeper understanding of the internal processes occurring within the fuel cells, namely electrochemical reactions, coupled to mass and heat transport phenomena. Most of these processes are difficult to realize experimentally, and thus mathematical models can be used to develop optimal control and operating strategies. Different types of approaches, such as analytical, semi-empirical or mechanistic models, have been followed by researchers. Analytical models are based on several simplifying assumptions in order to develop an appropriate correlation between analytical voltage and current density; nevertheless they are adequate tools to understand the effect of basic variables on the fuel cell performance. From another perspective, semi-empirical models are based on simple empirical equations to predict the fuel cell performance as a function of different operating conditions, such as pressure, temperature or fuel concentration. Finally, mechanistic models are transport models that use differential and algebraic equations, which are based on the electrochemistry and physics governing the phenomena taking place in the cell. These models generally employ numerical methods, and several approaches with different degrees of complexity are available in the literature. In Table 1 is presented a list of relevant numerical studies regarding two-phase flow phenomena in fuel cells: DMFC's and PEMFC's, considering both macro and micro scales.

Yang et al. [21] used a self-written code based on the SIMPLE algorithm with the finite-volume-method to develop a two-dimensional (2D) and isothermal model describing two-phase mass transport in liquid-feed DMFCs. The two-phase flow in the flow channels was described by the drift-flux model and the homogeneous mist-flow model concerning the anode and cathode respectively, and a micro-agglomerate model was developed for the cathode CL. Additionally, the model accounted for the effect of methanol and water crossover through the membrane, and was used to investigate the effects of different operating and structural parameters on the cell performance [21]. In order to reduce the numerical complexity related to the finite-volume technique coupled to CFD, a concise mathematical model was developed by Yang et al. [22] to cover the two-phase flow and under-rib convection phenomena at the anode. The two-phase effect is neglected at the cathode, and the oxygen concentration is assumed linearly distributed along the gas diffusion layer. The two-phase flow at the anode and pressure-driven convection were included in the model for a realistic but fast computational calculation and analysis of the DMFC design, and good agreement was found between the simulated and experimental results [22]. More recently, Guo et al. [23] developed a dynamic 3D and two-phase model to investigate the mass transfer characteristics in the air-breathing cathode of a DMFC. The model is based on general simplifications such as all the parameters are time-independent, the cathode GDL and CL are isotropic and homogeneous porous media, all gases are ideal and all phases are continuous, the two-phase flow is laminar and the fuel cell is operated isothermally without evaporation or condensation [23]. The model was implemented on FLUENT, and the multiphase mixture model was employed coupled to a pressure-based segregated algorithm. The model was validated using literature data and the numerical simulation results showed to agree well with the experimental data [23].

On the subject of PEMFC modeling at macro-scale, Iranzo et al. [17] developed a steady-state three-dimensional (3D) model for a fuel cell with parallel and serpentine flow field bipolar plates, using the built-in PEMFC module from CFD software ANSYS Fluent. The fluid flow pressure drop in GDLs and CLs is mainly governed by flow in porous media, represented by a negative source in the momentum equations [17]. Additionally, volumetric sources were added to the energy equation, accounting for reaction heat generation at the cathode CL, Joule effect, and latent heat for phase change. The blocking of the porous media and the flooding of the reaction surface were modeled by multiplying the porosity and the active surface area by an adequate factor, as well as the species diffusivities, which were also corrected to account for the pore blockage. In spite of featuring a single calibration parameter, the model predicted the performance of the cell with good accuracy for a wide range of operating conditions and different bipolar plate designs, and a good agreement between experimental and numerical model results was observed [17]. Cordiner et al. [18] also used the Fluent CFD solver to implement a steady-state 3D GDL model that could describe the water-saturation distribution on PEM fuel cells. The model was based on the assumption of laminar flow regime, the anode GDL was treated as single-phase flow, the liquid water was only produced at the cathode side and the product water coming from the cathode catalyst layer side was in vapor form [18]. ANSI-C written user defined functions (UDFs) were used to incorporate liquid water transport model into the GDL by means of a decoupled finite difference model. The authors stated that the use of this multiphase model allows to better evaluate the homogeneity of active area utilization, crucial for flow field and GDL design issues [18]. Mancusi et al. [40] studied the two-phase flow in PEMFC straight and tapered channels using the VOF method. The authors used a 3D model for simulate the fuel cell performance for diverse taper angles and temperatures and to calculate the total amount of water produced. This information was used as boundary conditions to simulate the two-phase flow in the cell through a 2D VOF model. The results reveal that water removal was enhanced tapering the channel downstream because of higher airflow velocity. In the straight channel the dominant pattern for water flow was slug flow, while in the tapered channel was film flow. Different contact angles were tested and it was concluded that as the GDL hydrophobicity is decreased the occurrence of film flow is more marked. More recently, Ferreira et al. [39] studied the two-phase flow in a straight anode channel of a PEMFC using the VOF method. The authors analyzed the influence of the hydrogen inlet velocity, operating temperature and channel walls wettability on water dynamics. In a hydrophilic channel, the water moves as films at the upper surface of channel while it moves as a droplet in a hydrophobic channel. The water is easily removed when using higher hydrogen inlet velocity, operating temperature and channel walls wettability.

Three dimensional simulations with CFD were also performed to investigate two-phase flow patterns in micro-fuel cells. Hsieh et al. [3] investigated the flow and heat transfer regarding developing laminar flows in μDMFCs with serpentine flow fields, and demonstrated that CFD could be employed in micro-fluidics with liquid as working medium considering Reynolds number (Re)≤421. The numerical simulation was successfully performed, and the friction factors, as well as the corresponding Nusselt numbers were calculated. Additionally, detailed velocity and temperature distributions in sharp U-bend regions were obtained in this study. Zhang et al. [4] also developed a 3D mathematical model for the two-phase flow in an air-breathing μDMFC. The authors employed finite-element analysis CFD to investigate the performance of a perforated cathode structure with parallel flow channels, assuming steady state and isothermal conditions, and neglecting the presence of water vapor in the cathode [4]. The simulation results indicated that the perforated structure with parallel flow channels could enhance the oxygen mass transport and reduce the water flooding to improve the cell performance, and the test results proved the validity of the simulated model [4]. More recently, Falcão et al. [41] used a CFD model with the VOF method to simulate the two-phase flow CO_2/methanol solution taking place at the anode side of a μDMFC. Simulations considering the carbon dioxide bubble profiles were performed and the influence of methanol solution and carbon dioxide inlet velocity and operating temperature were investigated. The model can be used to visualize in the entire serpentine channel the carbon dioxide profile and identifying zones where more bubbles are produced. The work by Fei et al. [25] presents numerical investigations of the two-phase flow in the anode micro-channels of DMFCs using the thermal lattice Boltzmann model (TLBM). The authors derived a 2D efficient and effective computational scheme, which was then examined by a commercial CFD package using the Navier-Stokes coupled to the volume-of-fluid (VOF) method. A single bubble is firstly generated at the micro-channel in all simulation cases. The upper and bottom walls are assumed to be no-slip, and the inflow and outflow boundary conditions are assigned to the inlet and exit of the simulation domain. The thermal effects on the bubble transport phenomena are investigated by varying the wall temperatures. The authors verified that an imposed positive wall temperature gradient is favorable for the bubble transport in the micro-channel, and that CO_2 bubbles can be successfully removed if the temperature gradient technique is applied in the bubbly flow [25]. Hutzenlaub et al. [27] investigated the bubble dynamics in the micro-channel of a DMFC taking into account the difference in wettability between the channel wall and the diffusion layer. The authors developed a dynamic 3D two-phase CFD/analytical model which is assumed to be isothermal, with laminar and incompressible flow. The contact angle hysteresis is fully developed, and the bubble shape is assumed to be constant and independent of transport velocity. Furthermore, the bubble menisci are treated as slip boundaries, while the channel walls are defined to be no-slip walls [27]. In order to reduce the calculation time and to allow for parametric studies, the model is further simplified by assuming constant pressure distribution in the gas within the bubble, and the capillary pressure drop over the bubble only occurs in the bypasses along the bubble. The model was validated by using experimentally measured bubble velocity, and excellent agreement was achieved [27]. Buie et al. [24] attempted to apply several existent two-phase flow models to their case study concerning a miniature liquid-fed DMFCs, and verified that pressure drop in micro-channels exhibited a different behavior from the previously explored DMFCs' anodes. The authors [24] performed a detailed parametric study of the hydrodynamic coupling between fuel flow and fuel cell operation by measuring the pressure drop, the number of gas slugs exiting the anode, and the slug velocity at the channel outlet. The conservation of momentum equation was used for the liquid phase, assumed to travel in an inertial frame of reference moving at the speed of the gas slugs, and the effect of the inertial forces was considered [24]. The results suggested that the pressure drop was dominated by interfacial forces, and the pressure data was sufficiently well characterized by a dual exponential fit. Moreover, by using empirical correlations for the number of gas slugs, the authors could fairly predict the pressure drop in the anode as a function of the current density [24]. Regarding PEMFC modeling at micro-scale, Su et al. [26] developed a 3D CFD model to investigate the local transport characteristics of a self-made μPEMFC, namely continuity, momentum and energy equations, species transport equation, liquid water saturation transport equation, and electrochemical reaction models. Micro-sensors, fabricated by micro-electro-mechanical systems (MEMS) technology, were adopted to measure the local temperature, voltage and current density. In this non-isothermal model, the Stefan-Maxwell equation was used to describe the species diffusion, the Butler-Volmer equation described the electrochemical reactions in the CLs, the Nernst-Planck equation described transport of protons through the membrane, and the Spring's model described the relationship between the membrane conductivity and its water content [26]. The good agreement between the experimental data and the predicted results suggests that the proposed CFD model can successfully assist in the design of a fuel cell [26]. Amara et al. [42] studied the droplet dynamics in a PEMFC micro-channel using lattice Boltzmann model approach. The authors simulated the density inside the micro-channel and observed that the droplet deformation increases for high capillary number values. The effect of wettability on droplet dynamics was also analyzed and it was concluded that a hydrophobic micro-channel enhance the droplets removal when compared to a hydrophilic one. Ding et al. [29, 30] investigated the effect of GDL microstructure on two-phase flow patterns by using the VOF method in a 3D isothermal numerical simulation of water droplets emerging from a GDL in micro-channels. The authors created a number of representative pores on the GDL surface, and different structures were investigated by changing the pore diameter and number of pore openings. Furthermore, the effect of GDL surface and channel wall surface wettability on the flow patterns was investigated by varying the water/surface static contact angles from 0° to 180°. The governing mass, momentum and volume fraction conservation equations were implemented using Fluent. The results suggested that the pressure drop is more dependent on the size of the droplets on the GDL surface than the level of water saturation, and for this reason empirical equations correlating pressure drop with water fraction must not be applied in fuel cell channels. Furthermore, it was found that at least 2 pores are required in the channel width direction to represent the microstructure of the GDL surface, and the two-phase flow pattern does not modify with an increase in the number of pores or a decrease in the pore diameter considering a constant GDL porosity. The same research group [30] further investigated the effect of the liquid injection rate and surface wettability on two-phase flow patterns by implementing a model with 320 pores on the GDL surface. The authors concluded that the use of highly hydrophilic materials helps to reduce the pressure drop, but also increases the water coverage ratio [30]. More recently, Chen et al. [31] developed a 3D dynamic two-phase model to investigate the liquid water behavior in the cathode of a PEMFC, and evaluate the effects of water distribution on the fuel cell mass transport and current density. In order to evaluate the dynamics of a single liquid water droplet, a larger pore is considered in the GDL, and the water generated by electrochemical reaction totally invades into that pore, thus entering the gas channel. To simplify the model, the fluid flow is considered to be isothermal, unsteady, laminar and incompressible, and either water evaporation or condensation is neglected [31]. VOF formulation is used to track the interface between the liquid water and air and UDFs written using C++ code were used for the effective diffusivity in the GDL and boundary conditions of oxygen mass fraction on the reactive surface [31]. The liquid water behavior in the gas channel was investigated for different inlet air velocities, wall wettability and water inlet positions. Based on the results obtained, the GDL structure was enhanced by increasing the pore number and arranging the pores near the gas channel side walls [31]. Kopanidis et al. [32] introduced the first approach considering a micro-pore scale geometry of a woven carbon cloth GDL in a 3D numerical simulation of flow and heat transfer in fuel cells. The goal was to find a connection between the pore scale geometry, the local heat and fluid flow and the condensation of water vapor that conduct to flooding. The numerical results were obtained regarding the channel pressure drop and convective heat transfer, and their effect on heat and mass transfer within the porous GDL was posteriorly evaluated [32]. Due to the local variation of the gas mixture quality, the values of mixture density, viscosity and specific heat are calculated based on the molecular weight and local humidity concentration. For sake of simplification, the current approach does not consider the effect of the bipolar plate on the temperature field. The presented numerical model is not able to provide simulations of the performance of a fuel cell, as several physical mechanisms are lacking such as anode geometry, membrane, bipolar plates and cooling system, larger serpentine channel geometry, electrochemical phenomena etc. However, this study has shown that the representation of the micro-pore scale structure combined with a CFD solution of a representative portion of the macro scale geometry is able to realistically simulate characteristics observed at macro-scale [32].

Table 1. Model approaches for two-phase flow dynamics in fuel cells

Reference	Scale	Type of Cell	Channel Geometry	Dimen-sions	State	Thermal condition	Software Used	Solution Method
Yang et al. 2007 [21]	Macro-scale	Active DMFC	n.a.	2D	Steady-state	Isothermal	Self-written computer code	SIMPLE algorithm + finite-volume-method
Yang et al. 2009 [22]	Macro-scale	Active DMFC	Single serpentine flow channel	2D	Steady-state	Isothermal	Self-written computer code	Complexity-reduced mathematical model
Guo et al. 2013 [23]	Macro-scale	Self-breathing DMFC	n.a.	3D	Transient	Isothermal	Fluent CFD package	Multiphase mixture model + PISO + PRESTO
Iranzo et al. 2010 [17]	Macro-scale	PEMFC	Serpentine and parallel flow channels	3D	Steady-state	Non-isothermal	ANSYS Fluent	Ansys build-in PEMFC module
Ferreira et al. 2015 [39]	Macro-scale	PEMFC	Single straight channel	3D	Transient	Isothermal	ANSYS Fluent 14.5.7	Build-in VOF method in Fluent
Mancusi et al. 2014 [40]	Macro-scale	PEMFC	Single straight channel and tapered flow channel	3D	Transient	Isothermal	Fluent 6.3 CFD package	Build-in VOF in Fluent+interface reconstruction algorithm
Cordiner et al. 2011 [18]	Macro-scale	PEMFC	n.a.	3D	Steady-state	Isothermal	Fluent 6.3 CFD package	ANSI-C written user defined functions + finite-volume CDF code Fluent 6.3

Reference	Scale	Type of Cell	Channel Geometry	Dimen-sions	State	Thermal condition	Software Used	Solution Method
Fei et al. 2010 [25]	Micro-scale	Active DMFC	Single straight micro-channel	2D	Transient	Non-isothermal	Commercial CFD package	Thermal lattice Boltzmann model + VOF
Hsieh et al. 2007 [3]	Micro-scale	Active DMFC	Serpentine flow channel	3D	Steady-state	Non-isothermal	Fluent 6.1 CFD package	Continuity, momentum, and energy equations solved by a general CFD code
Hutzenlaub et al. 2011 [27]	Micro-scale	Active DMFC	Single straight micro-channel	3D	Transient	Isothermal	CFD-ACE + simulation software ESI group	CFD + analytical model
Falcão et al. 2016 [41]	Micro-scale	Active DMFC	Serpentine flow channels	3D	Transient	Isothermal	ANSYS Fluent 14.5	Build-in VOF method in Fluent
Zhang et al. 2011 [4]	Micro-scale	Self-breathing DMFC	Perforated cathode structure with parallel channels	3D	Steady-state	Isothermal	COMSOL Multiphysics 3.6 Computation	Finite-element-method CFD code

Table 1. (Continued)

Reference	Scale	Type of Cell	Channel Geometry	Dimen-sions	State	Thermal condition	Software Used	Solution Method
Amara et al. 2015[42]	Micro-scale	PEMFC	Single straight micro-channel	3D	Transient	Isothermal	Commercial CFD package	lattice Boltzmann model based on the Shan-Chen Pseudo-potential model
Su et al. 2012 [26]	Micro-scale	PEMFC	Two-parallel serpentine flow channels	3D	Steady-state	Non-isothermal	CFD code-CFDRC	SIMPLE algorithm + finite-difference-method
Ding et al. 2010/2011 [29, 30]	Micro-scale	PEMFC	Single straight micro-channel	3D	Transient	Isothermal	Fluent 6.3.26 CFD package	Build-in VOF method in Fluent
Chen et al. 2012 [31]	Micro-scale	PEMFC	Single straight micro-channel	3D	Transient	Isothermal	Gambit 2.3 + Fluent 6.3.26 CFD package	C++ written user defined functions + VOF
Kopanidis et al. 2011 [32]	Micro-scale	PEMFC	Micro/pore scale geometry	3D	Steady-state	Non-isothermal	Wisetex software + CAD software + commercial CFD package	SIMPLE-based algorithm + finite-volume-method

Table 2. General characteristics of the image recording systems employed in two-phase flow visualization studies

Reference	Camera	Lens	Resolution	Frame Rate (fps)	Light
Lu et al. 2004 [43]	Sony digital video camera recorder + Nikon N70	MicroNikkor lens (60mm f/2.8 D)	n.a.	n.a.	n.a.
Yang et al. 2005 [44]	Sony DCR-TRV900E + JVC CCD	Navitar 6X CCD C-Mount Lens	n.a.	25	600 W spotlight Arrilite 600
Liao et al. 2007 [45]	Sony DCR-PC110E + Redlake MotionXtra HG-100K	n.a.	n.a.	30-125	Halogen spot lamp
Wong et al. 2005 [46]	Sony DCR-TRV900E + JVC CCD	Navitar 6X CCD C-Mount Lens	n.a.	n.a.	n.a.
Fu et al. 2007 [47]	Kodak motion coder SR-ultra	Undefined micro-lens	n.a.	500	250 W fiber optic illuminator
Kandlikar et al. 2009 [10]	Photron Fastcam-Ultima high-speed camera	Infinity model K2/STM long-distance microscope lens	1024×1024	60-2000	Dual light guide fiber optic light
Barreras et al. 2008, Lozano et al. 2008 [8, 9]	Hamamatsu ORCA-ER CCD camera	Nikon lens 50 mm f# 1.2	1024×1024	n.a.	Quantel YG781C-10 pulsed Nd:YAG laser
Yuan et al. 2016 [49]	Panasonic DMC-ZS10	n.a.	n.a.	n.a.	n.a.
Falcão et al. 2016[48]	Canon EOS 30D	n.a.	n.a.	n.a.	n.a.

Two-Phase Flow Visualization in Fuel Cells

The analysis of two-phase flow phenomena within μDMFCs provides a better understanding of the coherence of fluid motion, channel blockage, and cell performance, and although computational modeling is a valuable tool for the design and fabrication of micro-fuel cells, the feasibility of the computational simulation relies on the validation of the models with experimental data. The most basic method of experimental investigation is flow visualization, making possible the quantitatively or qualitatively investigation of the dynamic behavior of CO_2 gas bubbles and water droplets within fuel cells. Flow visualization techniques are generally based on image recording systems consisting of a charge coupled device (CCD), a mounted lens and a digital camcorder that registers the flow motion. Additionally, the working fluid may be altered in a way that the fluid motion stays unchanged, while the fluid transport is made detectable. Usually, the flow is made visible by using bright field illumination, where objects with a lower light transmission appear in the field as dark spots. However, few research works [10, 43-47] have been reported in visualization experiments regarding DMFCs, and particularly μDMFCs, due to the difficulty of implementing these techniques. In Table 2, the general characteristics of the image recording systems used for flow visualization by several authors are summarized.

Falcão et al. [48] performed visualization studies with a digital camera, at the anode side, in order to investigate the CO_2 bubbles formation and profile during the experiments with a passive micro DMFC. The pattern identified leading to better performances involves the formation of a large bubble at the superior left side of the cell and several small bubbles distributed through the active area. Yuan et al. [49] studied the two-phase flow of an active DMFC using the visualization method. Different flow fields and their effects on reactant and product managements were investigated. The visualization studies indicate that using a higher oxygen flow rate helps in water removal.

Lu et al. [43] visualized *in-situ* two-phase flow phenomena, and investigated the effects of backing pore structure and wettability on cell performance and two-phase flow dynamics. A digital video camera was used for the visualization of CO_2 bubbles in the anode side and water droplets in the cathode side through a transparent fuel cell, and the movie was then edited offline to capture motionless pictures of the two-phase flow. Yang et al. [44] reported a similar visualization study of the CO_2 bubble profile in the anode flow field of a transparent DMFC. Recurring to the reflection of the liquid-gas interface, the formation, growth, and departure of the bubbles from the surface of the GDL and their motion in the flow channel were distinctly captured by the image recording system. Furthermore, the same visualization method was employed by Liao et al. [45], who used a digital image recording system to report the experimental investigation of the dynamic behavior and visualization of the CO_2 gas bubbles in the anode channel of a DMFC. The motionless pictures were then captured from an offline edition of the movie.

A similar approach was used by Wong et al. [46] to investigate the CO_2 bubble profile regarding micro-scales in μDMFC. The authors [46] captured the flow pattern through a transparent fixture plate by using an image recording system, and the images were captured with a shutter speed of 1/250 s and time intervals of 1 second. Moreover, Fu et al. [47] also investigated the evolution of two-phase flow pattern and pressure drop in converging and diverging silicon-based micro-channels. By using a high-speed video camera, a micro-lens was mounted on the CCD to visualize the two-phase flow patterns in the diverging and converging micro-channels, and the projected area of the bubbles on the bottom wall in each frame was determined by manual edge detection [47]. Kandlikar et al. [10] employed a different approach, and monitored the fluid flow maldistribution in individual channels based on the entrance region pressure drop. The authors [10] validated the method by an *ex-situ* experimental setup simulating the two-phase flow in parallel channels, and an *in-situ* experimental setup with an operating fuel cell. In order to investigate the two-phase phenomena in the GDLs, a high-speed camera with a long-distance microscope lens was used to capture water formation and the flow patterns inside the channel.

Laser induced fluorescence techniques were also used to visualize the flow patterns within bipolar plates of fuel cells [8, 9]. In the studies performed by Barreras et al. [8], sulforhodamine B was used as luminescent tracer in a water-glycerin mixture in order to study velocity and pressure fields. The same research group investigated the gas passage across the GDLs [9] by seeding acetone vapor into the gaseous flow. This was achieved by bubbling the air inside an acetone bottle, and then the seeded acetone was excited by illumination with UV light. Its absorption peak for transitions to the first excited singlet state is located around 275 nm, while upon excitation acetone fluoresces in the blue, with a maximum intensity at a wavelength of 435 nm, a decay lifetime of 4 ns and an efficiency of 0.2%. The flow through the porous media (GDL) was described by using a physical sound coupling at the interface between the surface of the plate channels and the beginning of the backing layer [9]. Bazylak [13] provided a detailed overview on the visualization techniques applied for monitoring the water accumulation in conventional fuel cells.

Two-Phase Flow Visualization by Particle Image Velocimetry

The visualization methods can be quantitatively characterized recurring to image processing of the scalar distributions or using discrete particle distributions instead of scalar markers to visualize the flow. The most recognized technique of this category is Particle Image Velocimetry (PIV), an optical and non-intrusive method.

PIV is a technique that can be used to obtain instantaneous 2D velocity fields, providing detailed velocity data for comparison and validation of CFD models, it can also provide guidelines to improve the flow field designs of devices such as fuel cells. For these reasons, PIV has received increased attention in the last two decades, and many authors have employed PIV techniques to investigate two-phase flow dynamics in various regimes and types of flow channels [50-53], namely in fuel cells

[54-56]. A review on the developments of PIV for dispersed two-phase flow, and particularly for bubbly flows, is provided by Deen et al. [57].

PIV is used to determine the flow motion by measuring the displacement of passive tracer particles that are seeded into a fluid, given that the particles are small enough to closely follow the streamlines of the flow. A cross-section of the flow is illuminated by a laser light sheet, generally a double pulsed laser is used, which emits two pulses of light that cast two images of the particles. In most part of the cases, a CCD camera is used to record the images of the particles, capturing the two subsequent images separated by a short time interval caused by the time delay between pulses [55, 57]. The particles in the second image are thus slightly displaced from the first one, and a process known as interrogation is used to determine the displacement field between particles. In this process, each image is divided into a given number of cells, known as the interrogation areas, each of them containing a minimum amount of particles, and the mean displacement of the amount of particles contained in each interrogation area is then determined through a statistical correlation between the two images recorded [55, 57]. For traditional cross-correlation algorithms, a minimum of five particles per interrogation area is usually required, while for more advanced correlation algorithms this requirement may not be necessary [58]. When the number of particles is sufficient, the cross-correlation outcomes a dominant correlation peak within a background of noise peaks, which corresponds to the particle displacement. The velocity within the interrogation area can after that be easily determined by dividing the displacement by the image magnification and the time delay [57, 58].

A particularly critical aspect of PIV is the particle seeding, since for the most part of the cases the particles must be chemically inert, and they cannot be toxic, corrosive, abrasive, or volatile [59]. The large diversity of tracer materials used in PIV experiments, as well as the methods existent for particle generation, are summarized by Melling [59] for both liquid and gas flows. From the literature available, it is clear that the particle size determines their capability to correctly follow the flow [59-62]. As the size of the particles strongly influences the error of the measurement, monodisperse tracer particles are recommended to achieve uniform particle intensity.

However, due to the small size of the particles used for flow seeding, coupled to the strong velocity gradients that are developed between the walls, a particle migration perpendicular to the walls may occur. This is expected to occur at any Reynolds number, but the migration velocity depends on the effective value of Re. Roudet et al. [51] developed a methodology for PIV measurements concerning large field of view in planar confined geometries. Using a depth of field larger than the channel width, the in-plane variations of the velocity of the fluid averaged through the width of the channel were measured, and the operating conditions were investigated. The authors [51] observed that the tracer particles remain uniformly distributed considering low Re, whereas they all accumulate at the equilibrium position when large Re exist.

The two-phase also brings difficulties to the PIV measurements, which are inexistent in cases of single-phase flows. In many process industries, bubble column reactors are used to impose intensive mass transfer between the gases and liquids, since the gas bubbles should allow a large contact surface, and induce a local turbulence that enhances the mixing process [52]. However, the dispersed phase causes shadows that reduce the quality of the images, and the potential deformation of bubbles because the time delay can affect the precision of the measurement [57]. Numerous methods are available to discriminate and separate the information of the phases in the flow, such as the ensemble correlation, where the correlation functions of subsequent recordings of the flow are added up. In some two-phase flows, a short exposure time delay is required, therefore the ensemble correlation algorithm cannot be used [58]. Another approach is to detect the dispersed particles based on their size. Once the dispersed phase has been detected, two new image pairs are produced; in one pair the dispersed phase is masked and replaced with the average background intensity, at the same time as on the other pair the same is done for the continuous phase. At lower light intensities however, the distinction between tracer particles and bubbles is expected to be more difficult [57].

An experimental method to achieve simultaneous, phase-separated velocity measurements in two-phase flows was reported by Lindken et al. [52]. The proposed method is a combination of three PIV techniques: PIV with fluorescent tracer particles, shadowgraphy and digital masking. The shadowgraphy technique is based on the recording of a shadow image of the gas bubble, and the bubble velocity distribution is afterwards determined by a particle tracking velocimetry (PTV) algorithm. On the other hand, the digital masking technique is an operator in the PIV evaluation algorithm which avoids evaluation errors near the interface of the two faces. For the experiments, fluorescent tracer particles made of acryl mixed with fluorescent dye (Rhodamine 6G) were used, with average diameter of 10 μm. The influence of the contamination with tracers on the bubble rise velocity was earlier determined in other experiment [52]. Aiming at the mechanisms of small bubble distribution and their influence on the liquid turbulence, Pang et al. [53] introduced and compared three phase-discrimination methods between bubbles and tracer particles. In order to measure the hydrodynamics of bubbly flows in a 2D vertical channel, a PIV camera was used to evaluate at the same time the liquid turbulence and the bubble distribution. In the first method, a filter was applied to remove the tracer particles and noise from the original background images, and Sobel filters were applied to detect bubble edges independently of their direction. The identification of the tracer particles was based on a combination of Gaussian and Laplacian filters to optimize the raw images. The bubble contours were then subtracted from the image, and the resulting image contained only the tracer particles prepared for PIV calculation. In the second method, the bubble discrimination was performed based on a series of digital image filters including erosion filters, remove outliers, Prewit filters and median filters. Once the bubbles were identified, the image containing only the tracer particles was obtained by subtracting the bubble image from the raw image. Finally, the third method was based on the characteristic differences between the bubbles and the tracer particles, and their identification was performed manually in the raw images by using image-processing methods [53].

PARTICLE IMAGE VELOCIMETRY FOR FLOW VISUALIZATION IN FUEL CELLS

The PIV methodology was also applied to improve the knowledge and characterization of the fluid dynamics within the reactant transport channels of fuel cells. Most recently, Lindken and Burgmann [63] made a detailed review of laser-optical velocimetry methods that focus specifically on fuel cells.

Martin et al. [54] used PIV to investigate Re numbers in the range of 109–872 in a U-shaped fuel cell flow channel. These are typical Re numbers characterizing flow rates in fuel cells working from low to medium current densities. The PIV system was positioned to obtain data from two major planes of the flow, and one hundred pairs of images were acquired at each flow rate with time intervals ranging from 0.297–3 s. Vector post-processing was used to remove spurious vectors within the flow field. The authors [54] observed that for lower flow rates (Re<381), the instantaneous images of the U-bends of the channels are identical to the time-averaged flow representations, suggesting steady flow conditions. As the Re increases the fluctuations become more pronounced, and recirculation occurs at the outer corners of the U-bends. Feser et al. [55] used PIV to investigate the convective flow in different fuel cell designs, observing the movement of liquid water as opposed to the gaseous reactant species. The authors [55] measured 2D velocity fields in representative regions of interest within flow channels of interdigitated and single-serpentine PEM fuel cells, and verified that the porous medium can have a local variation in thickness or in permeability. However, since the analytical efforts developed by the authors have not considered the inertial effects, they showed to be incapable of predicting the true profile of the flow [55].

Lebaek et al. [56] investigated the influence of the inlet conditions on the manifold flow in a fuel cell stack using numerical (CFD) and experimental (PIV) methods. The pressure drop caused by the channels of the bipolar plate is simulated in both cases by an induced pressure drop associated to the resistance of bipolar plate channels. However, the test setup did not consider reactions, heat transfer, or two-phase flow phenomena within the fuel cell. A 2D PIV configuration was used for the experimental study, consisting of a Dantec PIV system, with a double-pulsed laser, a PIV camera and a Dantec Flowmap processor. The seeding was performed by a smoke generator and based on paraffin oil, and the droplets were about 1.55 μm, since the determined Stokes number suggested that these particles were able to efficiently respond to the flow. By using PIV, the authors could gather detailed information of the manifold flow and tiny segments of the field, however the PIV data failed to capture the near wall boundary layers, mainly due to the light reflection from the laser induced by the polished Plexiglas and oil film created by the seeding particles [56]. On the other hand, the PIV results allowed to obtain detailed information, exceeding the resolution of the mesh used in the actual CFD simulation, and thus the application of PIV allows the detailed understanding of the flow characteristics and of the overall fuel cell performance [56].

PARTICLE IMAGE VELOCIMETRY AT MICRO-SCALES

As a result of the miniaturization process, the macro-scale flow theory can no longer be applied to the flow through micro-channels. Experimental evidence suggests that fluid flow behavior in micro-channels differs from that in macro-channels since the length scale is decreased, and phenomena that are usually neglected at the macro-scale hence become more relevant [36, 64]. In order to overcome the often inconsistent and contradictory results obtained from the lab-scale experiments, improvements in methods for direct optical measurement of transport phenomena on these scales have been made [33, 34, 36, 64].

Micro-scale flow visualization has crucial in the development of microfluidics, and direct flow visualization is very important for a deeper understanding of micro-flows, allowing the investigation of non-ideal behavior such as spatial and temporal gradients in surface and fluid properties, and providing data for computational investigations [36]. Advances in the existing methods for direct optical measurement of transport phenomena concerning micro scales have been attained and extensive reviews presented by Koo and Kleinstreuer [64], Sinton [36], Jayaraj et al. [34] and Aubin et al. [33] describe in detail the most recurring methods for micro-scale flow characterization. Additionally, Morini et al. [35] presented a critical review of the measurement techniques for the analysis of gas micro-flows.

Micro-PIV techniques combine epi-fluorescence microscopy and specialized image interrogation algorithms, providing a robust method for measuring velocity fields in microfluidic systems. The micro-PIV was firstly introduced by Santiago et al. [60], and it has become the standard tool for fluid visualization and fluid velocity measurements in micro-channels. The details regarding its applications and guidelines can be found in Lindken et al. [37], Nguyen and Wereley [65] and Wereley and Meinhart [38].

The greatest advantage of this method is the intuitive vector plot format, where the vectors indicate the magnitude and direction of the velocity, allowing direct comparison with CFD simulations. Moreover, the method is flexible and allows the investigation of regions from only some micrometers up to various millimeters, as well as velocities from nanometers per second to meters per second, only by adapting the appropriate magnification or timing of the measurement devices [37, 38, 65].

Similar to PIV, for the optical measurement a transparent channel and optical access to the area of investigation are required. The fluid under study is seeded with tracer particles, and the particles must be small enough to adequately follow the flow, or their density should match the density of the fluid. Considering microfluidic applications working with water, the tracer particles are typically made of polystyrene and have diameters of 200 nm to 2 μm [37]. The simplest micro-PIV recording can be obtained with a microscope including a lamp for continuous illumination, and a digital camera in video mode to record pairs of successive images. Typically, micro-PIV realizations consist of a double-pulse light source, that is usually synchronized with the digital camera by a timing unit, where the time interval is defined by the time interval between the two synchronized light pulses [37]. Opposed to what occurs in PIV, instead of an area illuminated by a laser light sheet, the total flow field volume is illuminated and observed through a microscope objective. Although all the tracer particles emit light, only the light originating from the particles in the focus plane of the microscope objective, and of those slightly out-of-focus, is collected by the objective lens [37, 65].

On a typical epi-fluorescent micro-PIV system, a double-pulsed laser is used to generate monochromatic neodymium-doped yttrium aluminum garnet (Nd:YAG) laser light. Specifically designed for PIV systems, these lasers consist on two Nd:YAG laser cavities, beam-combining optics, and frequency doubling crystals. They emit two pulses of light, generally at a wavelength of 532 nm. For micro-PIV applications, less than 10 mJ of laser power per pulse is sufficient. In cases when continuous-chromatic arc lamps or halogen lamps are used instead of pulsed-light sources, an excitation filter must be applied such that only a narrow wavelength band of light illuminates the test section [37, 38]. The image formation of a micro-PIV recording is usually dominated by large magnification and high numerical aperture (NA) of the microscope objective.

For a higher image quality, the use of double-shutter cameras minimize the time interval between two consecutive recordings up to 1 μs, allowing measurements of high velocities at high magnifications. A CCD camera is typically used to record the particle-image fields. As the entire depth of the test section is illuminated, the measurement plane must be defined by the depth-of-field of the recording lens. This is feasible at micro-scale since the microscope objective lenses have large apertures with sharply defined objective planes, thereby allowing particles to transition quickly from being in focus to being out-of-focus [38]. The two subsequent digital images of the particles are stored in a computer, together with the time information, and the data is then evaluated by means of cross-correlation.

A rule usually adopted by the PIV community consists in selecting interrogation areas that simultaneously guarantee that at least five particle pairs are present, and that the displacement of the particles does not exceed a forth part of the interrogation area length. Additionally, if the velocity vector location within the interrogation area is not accurate, the use of sub-pixel interpolation schemes allows to estimate the average velocity [66]. The spatial resolution can be enhanced by imaging the particles with high numerical aperture, diffraction-limited optics and with sufficiently high magnification so that the particles are resolved with at least 3–4 pixels per particle diameter [38]. A review of the magnifications and NAs used in several micro-PIV investigations available in the literature is presented in Table 3.

The accuracy of the micro-PIV measurements is still compromised by technical issues, since it depends on the diffraction limit of the recording optics, the noise of the particle-image field and the interaction of the fluid with the finite-sized seed particles. Some studies concerning spatial resolutions of a few microns require that the particles are smaller than the laser light wavelength, and in these cases light scattering techniques cannot be employed. Also, particles with diameters minor than 1 μm are affected by the Brownian motion, which could compromise the evaluation of the flow velocity, since an additional uncertainty is introduced in the final results [37, 38, 65, 66].

Table 3. Parameters used in typical micro-PIV visualizations

Magnification (X)	NA	Channel Dimension (μm)	Reference
2.5	0.75	300×300	Hecht et al. 2012 [67]
5	0.15	1000×1000	Burgmann et al. 2013 [68]
5	0.10	200×200	Hsieh et al. 2008 [69]
10	0.30	516	Park et al. 2004 [70]
10	0.25	400×400	Fu et al. 2011, 2012 [71, 72]
10	0.20	780×260	Malsch et al. 2008 [73]
20	0.75	300×45	Lima et al. 2006 [74]
20	0.40	600	Klank et al. 2002 [75]
40	0.75	99	Park et al. 2004 [70]
40	0.60	200×115	Hsieh et al. 2004 [76]
60	1.40	300×30	Meinhart et al. 1999 [77]
100	1.40	n.a.	Santiago et al. 1998 [60]
5 **10** **20**	0.12 0.30 0.50	400×150	Gunther et al. 2004 [78]

The resolution of micro-PIV measurements can be improved using a super-resolution PIV, where the correlation analysis is primarily used to determine the average velocity of groups of particles in single interrogation areas, and then this information is fed into a particle tracking velocimetry algorithm to determine the displacement of single particles. This procedure can be developed to give ensemble-averaged velocity fields with spatial resolutions approaching one micron [60].

In order to reduce the influence of aspects such as the Brownian motion, low seed particle concentration and low-quality images, the ensemble-averaging technique was originally demonstrated by Meinhart et al. [58]. In this method, an ensemble of particle-image pairs is collected and each set of images is then spatially correlated. The resulting correlation functions are afterwards summed up and the peak in the correlation is determined from the sum of the correlations. The authors [58] compared three different algorithms applied to a series of images acquired from a steady Stokes flow, and verified that the correlation averaging increases the effective particle-image density. This allows the corresponding interrogation area to be reduced in size, thereby increasing the spatial resolution and evaluation fidelity, especially for low particle densities or high Brownian motions. As an alternative, Moghtaderi et al. [79] avoided the effect of the Brownian motion by using 700 nm red fluorescent polystyrene micro-spheres as seeding particles, when applying velocities of about 0.3 m s^{-1} in their experiments concerning the enhancement of the mixing behavior of reactant streams in a miniaturized hydrogen reactor.

The signal to noise ratio of the measurement can also be improved by applying optical filters in the microscope, placed between the light collecting objective and the digital camera. These barrier filters reflect the light at the illumination wavelength and transmit the light with longer wavelength. Thus, the light emitted from the tracer particles is collected on the camera sensor, while the other is blocked [37, 38]. It is desirable to use high quality optical filters with high transmission in the emission wavelength range, and the use of an inverted microscope system is convenient in most cases. Moreover, high particle concentration should be avoided, as higher background noise may occur due to the out-of-focus particles. In order to avoid the light contribution from the out-of-focus particles to the correlation signal, Silva et al. [66, 80] defined the depth-of-correlation as twice as the distance that a particle can be positioned from the object plane. Beyond that distance, the intensity of the particle is considered to be sufficiently low and cannot influence the velocity measurements. Other unexpected problems can affect the micro-PIV measurements, for instance the laser light can cause speckles that disturb the image quality, and thus that may be removed by using a holographic light-shaping diffuser plate in the beam path. Moreover, some authors [37] suggest the use of a plastic holographic diffuser plate to protect the optics from laser damage, in case the laser power is accidentally set too high. Progresses have also been made in order to deal with certain limitations of this technique, such as the incomplete characterization of the flow regarding 3D velocity fields. The development of the stereoscopic micro-PIV, which employs two cameras placed at an angle to one another and a stereomicroscope, permits to measure at the same time all three velocity components on the plane, thus allowing the full volume mapping form micro-PIV measurements [33, 37, 75].

Particle Image Velocimetry Measurements in Micro-Fuel Cells

Numerous publications report the use of micro-PIV for the study of several micro-fluidic devices [66, 69-90]. However, few researchers [61, 62, 67, 68, 91-95] have used this technique to visualize two-phase flows in micro-channels of fuel cells, mainly due to the difficulties found when seeding gas flows.

The application of micro-PIV on two-phase flows in micro-fluidic devices is the same as that used for single-phase flow, however the velocities are usually determined in one of the phases, typically in the liquid phase. The use of micro-PIV in liquids in a fuel cell related context was firstly published by Minor et al. [95]. The authors premixed a water droplet with fluorescent particles, and artificially injected it into a micro-channel with a real GDL material, in order to investigate the droplet deformation and contact angle hysteresis [95]. Fries et al. [85] investigated meandering micro-channels with a rectangular cross-section by using micro-PIV technology. These type of channel geometry, is commonly used for flow fields in fuel cells, and thus the flow characterization within meandering channels is of great importance for these micro-devices. The authors [85] seeded the liquid phase with Rhodamine-B, and the visualization of the flow was performed via an inverted fluorescence microscope equipped with a Rhodamine-B filter set and a dual-frame low noise CCD camera. A pulsed laser was used for illumination, and a commercial timing unit was used to synchronize the laser shots and the camera frames with a known time delay between frames. The vector field was calculated by a cross-correlation algorithm of subsequent paired micrographs and 200 micrographs were averaged for statistical reasons. In order to calculate the vector plots, vorticity values, swirling strength and streamlines of the flow, the DaVis 6 commercial software was used [85]. Interesting studies have also been performed by the research group of Weinmueller et al. [96, 97], who investigated the two-phase flow in micro-channels of fuel cells by using a fluorescence microscopy technique. The micro-sized channel networks were fabricated by a standard micro-electro-mechanical systems (MEMS) technique based on photolithography, and the microfluidic network on the chip has been designed in accordance to representative micro-reactors, for the applicability of micro fuel cell designs [96]. The methanol solution was doped with a Rhodamine 610 Chloride solution (Rhodamine B), a fluorine dye inert to the chemical oxidation of methanol. By using a laser with a pulse length of 10 ns as light source, the Rhodamine B re-emits light with a wavelength of around 594 nm, and thus the fluorescent liquid becomes clearly distinguishable from the non-luminous gas phase [96, 97]. The re-emitted light reaches the CCD camera yielding distinct two-phase flow images. Considering a single pair of gas-liquid mixture within a given channel, the flow regime can be constructed based on the different gas and liquid volume flow rates [96, 97]. The flow-visualization in the micro-channels illustrates the strong effect of CO_2 gas generation on the fluid flow and its strong influence on the homogeneous reactant supply to the membrane electrode assembly [97]. Most recently, the two-phase flow phenomena on the performance of an operating μDMFC was investigated by the research group of Burgmann et al. [68, 92, 93]. The authors used the micro-PIV technique in order to qualitatively and quantitatively investigate the appearance and evolution of CO_2 bubbles in the anode flow of a μDMFC [92, 93]. FluoRed tracer particles were added to a pre-heated water-methanol flow to obtain fluorescence, and these particles were then illuminated while their movement was detected by using a microscope. This way, the CO_2 bubble growth, as well as the transport of the bubbles, was analyzed for different volume flow rates of the methanol-water mixture. The micro-PIV set-up consisted of an epi-fluorescence microscope with a SensiCam and a frequency doubled laser, and the μDMFC was mounted on the microscope in order to allow the measurements [93]. A dichroic mirror within the light path was used to separate the particle light from the background light, and the measurement plane was defined by the focal plane of the microscope lens. Additionally to the micro-PIV measurements, a time resolved recording of the cell-voltage was performed [92, 93]. The authors stated that the cell power increased when big bubbles reduced the free cross-section area of the channel, since this would convectively force the methanol into the catalyst layer where it would be oxidized to form CO_2 [92, 93].

Micro-PIV studies to investigate *in situ* the flow field of the gas-phase in fuel cell-like micro-channels were performed by Yoon et al. [62] and Sugii and Okamoto [94]. However, the seeding particles used in their studies affected the performance of real operating fuel cells. Sugii and Okamoto [94] developed a fluorescent olive oil particle generator to optimize characteristics of the particles, such as diameter, fluorescent intensity, density, concentration, among others, and applied the technique to a fuel cell under non-operative conditions. Although the technique was useful for the investigation on flow fields of fuel cells with chemical reaction, the oil particles would be adsorbed in the GDLs disturbing the cell performance [68]. On the other hand, Yoon et al. [62] used two types of seeding particles, namely smoke particles and water droplets, and the flow delivery system had two possible paths for the air flow depending of the desired seeding. Either a smoke generator or a mist generator was included in the flow loop, as proper, and the flow was heated to reproduce conditions similar to that in an operational cell. The authors [62] verified that the Stokes number provided a measure of the accuracy of the particle to track the flow field, and performed a series of numerical fluid velocity and particle trajectory simulations, concluding that particles smaller than 1 μm are needed to accurately follow the flow. Moreover, the authors [62] encountered additional limitations when using these seeding particles, as the water droplets caused significant decelerations in the flow, and the smoke particles would damage the membrane decreasing the cell performance.

Other researchers tried different approaches in order to study gas flows in fuel cells. Hecht et al. [67, 91] coupled two techniques, namely micro-PIV and flow tagging, to study the flow and transport phenomena in the flow fields of fuel cells. In these studies, the cathode gas flow was replaced by a liquid, and the fluid water was seeded with 2 μm diameter Rhodamine B-labeled polystyrene particles that were imaged by micro-PIV. Alternatively, the global transport process was observed as a function of time using flow tagging, which allowed to visualize the propagating and mixing gas stream [67, 91]. In the flow tagging approach used in this work, the authors [67, 91] added nitrogen monoxide (NO) to the fresh gases by suddenly switching from an NO-free to a NO containing argon flow. Aiming to estimate the comparability of both measuring methods, the molecular flow tagging experiments were repeated with water as working fluid and Rhodamine B was used to visualize the flow and the mixing [67, 91]. The flow scene was illuminated by a frequency doubled Nd:YAG laser with wavelength equal to 532 nm in volume illumination mode, and it was further observed with a long pass filter of wavelength higher than 545 nm, and a CCD camera at 5 Hz frame rate. The measurements showed that the combination of the porous GDL and the serpentine channel structure lead to a pressure-driven transport through the gap between neighboring channels, and thus the transport is dominated by convection [67, 91].

Although this method provides important information about the gas distribution within fuel cells, it is based on the principle of Re-number similarity, which states that the dynamics of liquid and gas flows with the same Re are supposed to be equivalent [67, 91]. Thus fluid dynamic conditions must be achieved by Re-number scaling.

The research group of Burgmann et al. [61] presented an alternative gas seeding for micro-PIV in fuel cell-like micro-channels. In order to identify appropriate particles for wall-bounded gas flows, numerous materials and different types of particle generators were examined to evaluate the particle size distributions and concentrations, and to investigate the capability of the generated particles to follow the flow. Typical fluorescent dyes such as Rhodamine-B, Nile Blue A or Ethidium Bromide are the most commonly used, especially Rhodamine-B which is easily dissolved in various liquids that generate aerosols for PIV-measurements. Nevertheless, for methanol there is a critical concentration that determines the maximum intensity of fluorescence [61]. In this study, the authors [61] performed micro-PIV measurements and CFD calculations to examine the applicability of the seeding particles. The micro-PIV set-up consisted of an epi-fluorescence microscope, a Sensicam and a frequency doubled laser, and the DaVis 7.2 PIV software from LaVision was used. The authors used cross-correlation schemes with interrogation window overlap of 50% and a spatial resolution of 38 μm for the velocity data. Given the small density difference, the deviation of the particles was considered negligible. 3D-CFD calculations were performed using FLUENT, and the deviation of particles with different diameters was calculated. The authors concluded that atomizing ethylene-glycol with Rhodamine-B seemed to be suitable for micro-PIV, since those particles adequately follow the flow and emit sufficient fluorescent light. Moreover, recent measurements of an operating DMFC adding such a seeding to the cathode side flow proved that the cell performance is not altered by those particles [61, 68]. Based on their previous study [61], the research group of Burgmann et al. [68] presented the latest work published on two-phase flows regarding μDMCS. The authors developed a methodology to investigate synchronously the fluid mechanics and cell performance by using micro-PIV and a time-resolved voltage measurement. The water-methanol flow of the anode side was premixed with FluoRed polystyrol tracer particles with a diameter of 1.22 μm manufactured by micro-particles GmbH. The seeding density was considered sufficient for the application of cross-correlation schemes in 64×64 pixels interrogation areas, corresponding to more than 10 particles per interrogation area [68]. At the cathode side, the air flow was mixed with ethylene-glycol particles generated by a LaVision atomizer, with a mean particle size of approximately 1 μm. Crystalline Rhodamine-B was dissolved in ethylene-glycol such that the solubility limit was reached to get a fluorescence signal. However, the authors stated that the particle density within the micro-channel at the cathode was poor, of about 10–20 particles per image [68]. A single straight channel configuration was chosen for the DMFC design in this work, in order to avoid parasitic effects like flow shortcut, and the region of interest (ROI) of the micro-PIV measurements of the anode channel length was placed in its midsection. On the cathode side, the image quality was much inferior than in the anode, and thus more and stronger image correction and manipulation was needed to separate the particles from the background. Moreover, due to the necessity of providing optical access, the cell could not be compressed as usual, and for this reason, the cell performance obtained was quite poor [68]. Although the results provided improved knowledge on the effects of two-phase flow in the cell performance, the authors concluded that further investigation is needed, together with an electrically segmented cell voltage measurement technique.

Calabriso et al. [98] studied the bubbly flow mapping in a serpentine anode channel of a DMFC using PIV technique. The fluid motion is followed using a pearlescent silver powder particle tracer with 10-100 um (Timiron Super Silver) mixed with the water-methanol solution with a concentration of 2.5 mg/L. A high speed camera, Photron APX was placed in font o the anode, where the Plexigas support allows for images acquisition. N images (around 2000) are pre-processed with Matlab by using a minimum subtraction technique to diminish the background noise while the velocity field computation is done with PivLab using a cross-correlation algorithm. Statistical analysis, based on PIV acquisition, determined the mean bubble size and the mean residence time of the bubble in the channel, at different flow rates. Smaller bubbles can be more easily detached from the GDL and dragged along the channel by the liquid phase. The authors observed three regimes of Re number in a preliminary statistical analysis (80, 400, 800). For Re of 400 and 800 comparable behaviors were reported: smaller size of bubbles and smaller residence time with Re number increasing. At Re of 80 the same situation was observed for low current density but, for current densities higher than 35 mA/cm^2, a transition from bubbly to slug flow was recorded. The coalescence of bubbles can originate a decrease in power output, between 0.1 and 0.2 V, namely when the Re number is lower than 80 and the current density is over the 35 mA/cm^2.

CONCLUSION

A review was made regarding the existing visualization methods applied primarily for the determination of two-phase flow phenomena and flow velocity data in fuel cells at macro and micro-scales. Although many scalar-based visualization methods were used to investigate the CO_2 bubble flow effect in direct methanol fuel cells, the most established method of this kind is digital Particle Image Velocimetry. For this reason, the PIV methodology was widely applied to improve the understanding and characterization of the two-phase fluid dynamics within the reactant transport channels of DMFCs.

As a result of the miniaturization process of DMFCs, the occurrence of two-phase flow phenomena can lead to the obstruction of the active catalyst sites, oxidant starvation, and reduction of the mass transport velocity. Although micro-PIV is a well-established technique for the visualization of two-phase flow phenomena within various regimes and types of flow channels, this technique is not commonly applied in micro-fuel cell research mainly due to seeding problems, especially in micro-gas flows.

For this reason, future developments regarding micro-PIV application in micro-fuel cells are needed. Moreover, an adequate methodology for gas flows in micro-channels is still lacking, nevertheless some research groups are actively working on this issue and solutions for gas flows applications are expected over short term.

ACKNOWLEDGMENTS

D.S. Falcão acknowledges the post-doctoral fellowship (SFRH/BDP/109815/2015) supported by the Portuguese "Fundação para a Ciência e Tecnologia" (FCT), POPH/QREN and European Social Fund (ESF). J.P. Pereira acknowledges the project PTDC/EQU-FTT/112475/2009, supported by FCT. POCI (FEDER) also supported this work via CEFT (UID/EME/00532/2013).

REFERENCES

[1] S. K, A. P, L. V. *Mini-Micro Fuel Cells Fundamentals and Applications:* Springer; 2008.

[2] Sundarrajan S, Allakhverdiev SI, Ramakrishna S. Progress and perspectives in micro direct methanol fuel cell. *Int J Hydrogen Energy.* 2012;37(10):8765-86.

[3] Hsieh S-S, Her B-S. Heat transfer and pressure drop in serpentine μDMFC flow channels. *Int J Heat Mass Tran.* 2007 12//;50(25–26):5323-7.

[4] Zhang Y, Zhang P, He H, Zhang B, Yuan Z, Liu X, et al. A self-breathing metallic micro-direct methanol fuel cell with the improved cathode current collector. *International Journal of Hydrogen Energy.* 2011 1//;36(1):857-68.

[5] Wong CW, Zhao TS, Ye Q, Liu JG. Experimental investigations of the anode flow field of a micro direct methanol fuel cell. *J Power Sources.* 2006;155(2):291-6.

[6] Lu Y, Reddy RG. Effect of flow fields on the performance of micro-direct methanol fuel cells. *International Journal of Hydrogen Energy.* 2011 1//;36(1):822-9.

[7] Zhang Y, He H, Yuan Z, Wang S, Liu X. Effect of design and operating parameters on dynamic response of a micro direct methanol fuel cell. *International Journal of Hydrogen Energy.* 2011 2//;36(3):2230-6.

[8] Barreras F, Lozano A, Valiño L, Mustata R, Marín C. Fluid dynamics performance of different bipolar plates: Part I. Velocity and pressure fields. *J Power Sources.* 2008 1/10/;175(2):841-50.

[9] Lozano A, Valiño L, Barreras F, Mustata R. Fluid dynamics performance of different bipolar plates: Part II. Flow through the diffusion layer. *J Power Sources.* 2008 5/1/;179(2):711-22.

[10] Kandlikar SG, Lu Z, Domigan WE, White AD, Benedict MW. Measurement of flow maldistribution in parallel channels and its application to ex-situ and in-situ experiments in PEMFC water management studies. *Int J Heat Mass Tran.* 2009 3//;52(7–8):1741-52.

[11] Hidrovo CH, Kramer TA, Wang EN, Vigneron S, Steinbrenner JE, Koo JM, et al. Two-Phase Microfluidics for Semiconductor Circuits and Fuel Cells. *Heat Transfer Engineering.* 2006;27(4):53-63.

[12] Litster S, Sinton D, Djilali N. Ex situ visualization of liquid water transport in PEM fuel cell gas diffusion layers. *J Power Sources.* 2006 3/9/;154(1):95-105.

[13] Bazylak A. Liquid water visualization in PEM fuel cells: A review. *Int J Hydrogen Energy.* 2009 5//;34(9):3845-57.

[14] Gupta R, Fletcher DF, Haynes BS. CFD modelling of flow and heat transfer in the Taylor flow regime. *Chemical Engineering Science.* 2010 3/15/;65(6):2094-107.

[15] Asadolahi AN, Gupta R, Fletcher DF, Haynes BS. CFD approaches for the simulation of hydrodynamics and heat transfer in Taylor flow. *Chemical Engineering Science.* 2011 11/15/;66(22):5575-84.

[16] Asadolahi AN, Gupta R, Leung SSY, Fletcher DF, Haynes BS. Validation of a CFD model of Taylor flow hydrodynamics and heat transfer. *Chemical Engineering Science.* 2012 2/13/;69(1):541-52.

[17] Iranzo A, Muñoz M, Rosa F, Pino J. Numerical model for the performance prediction of a PEM fuel cell. Model results and experimental validation. *Int J Hydrogen Energy.* 2010 10//;35 (20):11533-50.

[18] Cordiner S, Lanzani SP, Mulone V. 3D effects of water-saturation distribution on polymeric electrolyte fuel cell (PEFC) performance. *Int J Hydrogen Energy.* 2011 8//;36(16):10366-75.

[19] Gostick JT, Ioannidis MA, Fowler MW, Pritzker MD. Pore network modeling of fibrous gas diffusion layers for polymer electrolyte membrane fuel cells. *J Power Sources.* 2007 11/8/;173(1):277-90.

[20] Lee K-J, Nam JH, Kim C-J. Pore-network analysis of two-phase water transport in gas diffusion layers of polymer electrolyte membrane fuel cells. *Electrochim Acta.* 2009 1/30/;54(4):1166-76.

[21] Yang WW, Zhao TS. A two-dimensional, two-phase mass transport model for liquid-feed DMFCs. *Electrochim Acta.* 2007 6/10/;52(20):6125-40.

[22] Yang Y, Liang YC. Modelling and analysis of a direct methanol fuel cell with under-rib mass transport and two-phase flow at the anode. *J Power Sources.* 2009 12/1/;194(2):712-29.

[23] Guo H, Chen YP, Xue YQ, Ye F, Ma CF. Three-dimensional transient modeling and analysis of two-phase mass transfer in air-breathing cathode of a fuel cell. *Int J Hydrogen Energy.* 2013 8/21/;38(25):11028-37.

[24] Buie CR, Santiago JG. Two-phase hydrodynamics in a miniature direct methanol fuel cell. *Int J Heat Mass Tran.* 2009 10//;52(21–22):5158-66.

[25] Fei K, Chen TS, Hong CW. Direct methanol fuel cell bubble transport simulations via thermal lattice Boltzmann and volume of fluid methods. *J Power Sources*. 2010 4/2/;195(7):1940-5.

[26] Su A, Ferng YM, Chen WT, Cheng CH, Weng FB, Lee CY. Investigating the transport characteristics and cell performance for a micro PEMFC through the micro sensors and CFD simulations. *Int J Hydrogen Energy*. 2012 8//;37(15):11321-33.

[27] Hutzenlaub T, Paust N, Zengerle R, Ziegler C. The effect of wetting properties on bubble dynamics and fuel distribution in the flow field of direct methanol fuel cells. *J Power Sources*. 2011 10/1/;196(19):8048-56.

[28] Wang S-J, Huo W-W, Zou Z-Q, Qiao Y-J, Yang H. Computational simulation and experimental evaluation on anodic flow field structures of micro direct methanol fuel cells. *Appl Therm Eng*. 2011 10//;31(14–15):2877-84.

[29] Ding Y, Bi HT, Wilkinson DP. Three-dimensional numerical simulation of water droplet emerging from a gas diffusion layer surface in micro-channels. *J Power Sources*. 2010 11/1/;195(21):7278-88.

[30] Ding Y, Bi HT, Wilkinson DP. Three dimensional numerical simulation of gas–liquid two-phase flow patterns in a polymer–electrolyte membrane fuel cells gas flow channel. J *Power Sources*. 2011 8/1/;196(15):6284-92.

[31] Chen L, Cao T-F, Li Z-H, He Y-L, Tao W-Q. Numerical investigation of liquid water distribution in the cathode side of proton exchange membrane fuel cell and its effects on cell performance. *Int J Hydrogen Energy*. 2012 6//;37(11):9155-70.

[32] Kopanidis A, Theodorakakos A, Gavaises M, Bouris D. Pore scale 3D modelling of heat and mass transfer in the gas diffusion layer and cathode channel of a PEM fuel cell. *Int J Therm Sci*. 2011 4//;50(4):456-67.

[33] Aubin J, Ferrando M, Jiricny V. Current methods for characterising mixing and flow in microchannels. *Chemical Engineering Science*. 2010 3/15/;65(6):2065-93.

[34] Jayaraj S, Kang S, Suh YK. A Review on the Analysis and Experiment of Fluid Flow and Mixing in Micro-Channels. *Journal of Mechanical Science and Technology*. 2007;21:536-48.

[35] Morini GL, Yang Y, Chalabi H, Lorenzini M. A critical review of the measurement techniques for the analysis of gas microflows through microchannels. *Experimental Thermal and Fluid Science*. 2011 9//;35(6):849-65.

[36] Sinton D. Microscale flow visualization. *Microfluid Nanofluid*. 2004;1:2-21.

[37] Lindken R, Rossi M, Grobe S, Westerweel J. Micro-Particle Image Velocimetry (mPIV): Recent developments, applications, and guidelines. *Lab Chip*. 2009;9:2551-67.

[38] Wereley ST, Meinhart CD. Recent Advances in Micro-Particle Image Velocimetry. *Annu Rev Fluid Mech*. 2010;42:557-76.

[39] Ferreira RB, Falcão DS, Oliveira VB, Pinto AMFR. Numerical simulations of two-phase flow in an anode gas channel of a proton exchange membrane fuel cell. *Energy*. 2015;82:619-28.

[40] Mancusi E, Fontana É, Ulson de Souza AA, Guelli Ulson de Souza SMA. Numerical study of two-phase flow patterns in the gas channel of PEM fuel cells with tapered flow field design. *Int J Hydrogen Energy*. 2014;39(5):2261-73.

[41] Falcão DS, Pereira JP, Pinto AMFR. Numerical simulations of anode two-phase flow in Micro-DMFC using the volume of fluid method. *Int J Hydrogen Energy*. 2016;41(43):19724-30.

[42] Ben Amara MEA, Ben Nasrallah S. Numerical simulation of droplet dynamics in a proton exchange membrane (PEMFC) fuel cell micro-channel. *Int J Hydrogen Energy*. 2015;40(2):1333-42.

[43] Lu GQ, Wang CY. Electrochemical and flow characterization of a direct methanol fuel cell. *J Power Sources*. 2004 7/12/;134(1):33-40.

[44] Yang H, Zhao TS, Ye Q. In situ visualization study of CO2 gas bubble behavior in DMFC anode flow fields. *J Power Sources*. 2005 1/4/;139(1–2):79-90.

[45] Liao Q, Zhu X, Zheng X, Ding Y. Visualization study on the dynamics of CO2 bubbles in anode channels and performance of a DMFC. *J Power Sources.* 2007 9/27/;171(2):644-51.

[46] Wong CW, Zhao TS, Ye Q, Liu JG. Transient Capillary Blocking in the Flow Field of a Micro-DMFC and Its Effect on Cell Performance. *Journal of the Electrochemical Society.* 2005;152(8):A1600-A5.

[47] Fu BR, Tseng FG, Pan C. Two-phase flow in converging and diverging microchannels with CO2 bubbles produced by chemical reactions. International Journal of Heat and Mass Transfer. 2007 1//;50(1–2):1-14.

[48] Falcão DS, Pereira JP, Pinto AMFR. Effect of stainless steel meshes on the performance of passive micro direct methanol fuel cells. *Int J Hydrogen Energy.* 2016;41(31):13859-67.

[49] Yuan W, Wang A, Yan Z, Tan Z, Tang Y, Xia H. Visualization of two-phase flow and temperature characteristics of an active liquid-feed direct methanol fuel cell with diverse flow fields. *Appl Energy.* 2016;179:85-98.

[50] Liu Y-P, Wang P-Y, Wang J, Du Z-H. Investigation of Taylor bubble wake structure in liquid nitrogen by PIV technique. *Cryogenics.* 2013 5//;55–56(0):20-9.

[51] Roudet M, Billet A-M, Risso F, Roig V. PIV with volume lighting in a narrow cell: An efficient method to measure large velocity fields of rapidly varying flows. *Experimental Thermal and Fluid Science.* 2011 9//;35(6):1030-7.

[52] Lindken R, Merzkirch W. A novel PIV technique for measurements in multiphase flows and its application to two-phase bubbly flows. Exp Fluids. 2002;33:814-25.

[53] Pang M, Wei J. Experimental investigation on the turbulence channel flow laden with small bubbles by PIV. Chemical Engineering Science. 2013 5/3/;94(0):302-15.

[54] Martin J, Oshkai P, Djilali N. Flow Structures in a U-Shaped Fuel Cell Flow Channel: Quantitative Visualization Using Particle Image

Velocimetry. *Journal of Fuel Cell Science and Technology.* 2004;2(1):70-80.

[55] Feser JP, Prasad AK, Advani SG. Particle Image Velocimetry Measurements in a Model Proton Exchange Membrane Fuel Cell. *Journal of Fuel Cell Science and Technology.* 2007;4:8.

[56] Lebaek J, Andreasen MB, Andresen HA, Bang M, Kaer SK. Particle Image Velocimetry and Computational Fluid Dynamics Analysis of Fuel Cell Manifold. *Journal of Fuel Cell Science and Technology.* 2010;7:10.

[57] Deen NG, Westerweel J, Delnoij E. Two-Phase PIV in Bubbly Flows: Status and Trends. *Chemical Engineering Technology.* 2002;25:97-101.

[58] Meinhart CD, Wereley ST, Santiago JG. A PIV Algorithm for Estimating Time-Averaged Velocity Fields. *Journal of Fluids Engineering.* 2000;122:285-9.

[59] Melling A. Tracer particles and seeding for particle image velocimetry. *Meas Sci Technology.* 1997;8:1406-16.

[60] Santiago JG, Wereley ST, Meinhart CD, Beebe DJ, Adrian RJ. A particle image velocimetry system for microfluidics. *Exp Fluids.* 1998;25:316-9.

[61] Burgmann S, Schoot N, Asbach C, Wartmann J, Lindken R. Analysis of tracer particle characteristics for micro PIV in wall-bounded gas flows. *La Houille Blanche.* 2011;4:55-61.

[62] Yoon SY, Ross JW, Mench MM, Sharp KV. Gas-phase particle image velocimetry (PIV) for application to the design of fuel cell reactant flow channels. *J Power Sources.* 2006 10/6/;160(2):1017-25.

[63] Lindken R, Burgmann S. Laser-optical methods for transport studies in low temperature fuel cells. In: Hartnig C RC, editor. *Polymer electrolyte membrane and direct methanol fuel cell technology,* Volume 2: In situ characterization techniques for low temperature fuel cells. Cambridge: Sawston; 2012. p. 425-61.

[64] Koo J, Kleinstreuer C. Liquid flow in microchannels: experimental observations and computational analyses of microfluidics effects. *Journal of micromechanics and microengineering.* 2003;13:568-79.

[65] Nguyen NT, Wereley ST. Fundamentals and Applications of Microfluidics. 2nd ed. ed. House A, editor. Boston2006.

[66] Silva G, Leal N, Semiao V. Determination of microchannels geometric parameters using micro-PIV. *Chemical Engineering Research and Design.* 2009 3//;87(3):298-306.

[67] Hecht C, Schoot N, Kronemayer H, Wlokas I, Lindken R, Schulz C. Visualization of the gas flow in fuel cell bipolar plates using molecular flow seeding and micro-particle image velocimetry. *Exp Fluids.* 2012;52:743-8.

[68] Burgmann S, Blank M, Panchenko O, Wartmann J. lPIV measurements of two-phase flows of an operated direct methanol fuel cell. *Exp Fluids.* 2013;54:1513:14.

[69] Hsieh S-S, Huang Y-C. Passive mixing in micro-channels with geometric variations through μPIV and μLIF measurements. *Journal of Micromechanics and Microengineering.* 2008;18:065017.

[70] Park JS, Choi CK, Kihm KD. Optically sliced micro-PIV using confocal laser scanning microscopy (CLSM). *Experiments in Fluids.* 2004;37:105-19.

[71] Fu T, Ma Y, Funfschilling D, Li HZ. Dynamics of bubble breakup in a microfluidic T-junction divergence. *Chemical Engineering Science.* 2011 9/15/;66(18):4184-95.

[72] Fu T, Wu Y, Ma Y, Li HZ. Droplet formation and breakup dynamics in microfluidic flow-focusing devices: From dripping to jetting. *Chemical Engineering Science.* 2012 12/24/;84(0):207-17.

[73] Malsch D, Kielpinski M, Merthan R, Albert J, Mayer G, Köhler JM, et al. μPIV-Analysis of Taylor flow in micro channels. *Chemical Engineering Journal.* 2008 1/15/;135, Supplement 1(0):S166-S72.

[74] Lima R, Wada S, Tsubota K, Yamaguchi T. Confocal micro-PIV measurements of three dimensional profiles of cell suspension flow in a square microchannel. *Measurement Science and Technology.* 2006;17:797-808.

[75] Klank H, Goranovic G, Kutter JP, Gjelstrup H, Michelsen J, Westergaard CH. PIV measurements in a microfluidic 3D-sheathing structure with three-dimensional flow behaviour. *Journal of Micromechanics and Microengineering.* 2002;12:862-9.

[76] Hsieh S-S, Lin C-Y, Huang CF, Tsai H-H. Liquid flow in a micro-channel. *Journal of micromechanics and microengineering.* 2004;14:436-45.

[77] Meinhart CD, Wereley ST, Santiago JG. PIV measurements of a microchannel flow. *Exp Fluids.* 1999;27:414-9.

[78] Gunther A, Khan SA, Thalmann M, Trachsel F, Jensen KF. Transport and reaction in microscale segmented gas–liquid flow. *Lab Chip.* 2004;4:278-86.

[79] Moghtaderi B, Shames I, Djenidi L. Microfluidic characteristics of a multi-holed baffle plate micro-reactor. *International Journal of Heat and Fluid Flow.* 2006 12//;27(6):1069-77.

[80] Silva G, Leal N, Semiao V. Micro-PIV and CFD characterization of flows in a microchannel: Velocity profiles, surface roughness and Poiseuille numbers. *International Journal of Heat and Fluid Flow.* 2008 8//;29(4):1211-20.

[81] Qu W, Mudawar I, Lee SY, Wereley ST. Experimental and Computational Investigation of Flow Development and Pressure Drop in a Rectangular Micro-channel. *Journal of Electronic Packaging.* 2006;128:1-9.

[82] Wu Z, Nguyen NT, Huang X. Nonlinear diffusive mixing in microchannels: theory and experiments. *Journal of micromechanics and microengineering.* 2004;14:604-11.

[83] Waelchli S, Rudolf von Rohr P. Two-phase flow characteristics in gas–liquid microreactors. *International Journal of Multiphase Flow.* 2006 7//;32(7):791-806.

[84] King C, Walsh E, Grimes R. PIV measurements of flow within plugs in a microchannel. *Microfluid Nanofluid.* 2007;3:463-72.

[85] Fries DM, Waelchli S, Rudolf von Rohr P. Gas–liquid two-phase flow in meandering microchannels. *Chemical Engineering Journal.* 2008 1/15/;135, Supplement 1(0):S37-S45.

[86] Dore V, Tsaoulidis D, Angeli P. Mixing patterns in water plugs during water/ionic liquid segmented flow in microchannels. *Chemical Engineering Science.* 2012 10/1/;80(0):334-41.

[87] Zaloha P, Kristal J, Jiricny V, Völkel N, Xuereb C, Aubin J. Characteristics of liquid slugs in gas–liquid Taylor flow in microchannels. *Chemical Engineering Science.* 2012 1/22/;68(1):640-9.

[88] Anastasiou AD, Makatsoris C, Gavriilidis A, Mouza AA. Application of μ-PIV for investigating liquid film characteristics in an open inclined microchannel. *Experimental Thermal and Fluid Science.* 2013 1//;44(0):90-9.

[89] Lima R, Wada S, Tanaka S, Takeda M, T. I, Tsubota K, et al. In vitro blood flow in a rectangular PDMS microchannel: experimental observations using a confocal micro-PIV system. *Biomed Microdevices.* 2007;10:153-67.

[90] Park JS, Kihm KD. Use of confocal laser scanning microscopy (CLSM) for depthwise resolved microscaleparticle image velocimetry. *Optics and Lasers in Engineering.* 2006;44:208-23.

[91] Hecht C, Schoot N, Kronemayer H, Lindken R, Schulz C, editors. Visualization of the gas flow in fuel cell bipolar plates using molecular flow tagging velocimetry and micro PIV. 15th Int Symp on Applications of Laser Techniques to Fluid Mechanics; 2010; Lisbon, Portugal.

[92] Blank M, Burgmann S, Wartmann J, editors. μPIV measurements of the CO2-bubble evolution in the anode flow of an operated direct methanol fuel cell. 16th Int Symp on Applications of Laser Techniques to Fluid Mechanics; 2012; Lisbon, Portugal.

[93] Burgmann S, Blank M, Wartmann J, Heinzel A. Investigation of the Effect of CO2 bubbles and Slugs on the Performance of a DMFC by Means of Laser-optical Flow Measurements. *Energy Procedia.* 2012;28(0):88-101.

[94] Sugii Y, Okamoto K, editors. Velocity measurement of gas flow using micro PIV technique in polymer electrolyte fuel cell. ASME

4th International Conference on Nanochannels, Microchannels, and Minichannels; 2006; Ireland: ASME.

[95] Minor G, Zhu X, Oshkai P, Sui P, Djilali N. Water transport dynamics in fuel cell micro-channels. In: Kakaç S, editor. *Mini-micro fuel cells: fundamentals and applications.* Berlin2008. p. 153-70.

[96] Weinmueller C, Hotz N, Mueller A, Poulikakos D. On two-phase flow patterns and transition criteria in aqueous methanol and CO2 mixtures in adiabatic, rectangular microchannels. *International Journal of Multiphase Flow.* 2009 8//;35(8):760-72.

[97] Weinmueller C, Tautschnig G, Hotz N, Poulikakos D. A flexible direct methanol micro-fuel cell based on a metalized, photosensitive polymer film. *J Power Sources.* 2010;195(12):3849-57.

[98] Calabriso A, Borello D, Romano GP, Cedola L, Del Zotto L, Santori SG. Bubbly flow mapping in the anode channel of a direct methanol fuel cell via PIV investigation. *Appl Energy.* 2017;185, Part 2:1245-55.

In: Direct Methanol Fuel Cells
Editors: Rose Hernandez et al.
ISBN: 978-1-53612-603-7

Chapter 4

RECENT DEVELOPMENTS IN PASSIVE DIRECT METHANOL FUEL CELLS

B. A. Braz, V. B. Oliveira[*] and A. M. F. R. Pinto[†]
CEFT (Transport Phenomena Research Center),
Department of Chemical Engineering,
Faculty of Engineering, University of Porto, Porto, Portugal

ABSTRACT

Direct methanol fuel cells (DMFCs) are in the Energy Agenda due to their market potential to replace the conventional batteries in portable applications, either for recreational, professional, military and medical purposes, since they enable the direct conversion of the chemical energy stored in a fuel, methanol, to electrical energy with water and carbon dioxide as final products. Moreover, DMFCs offer higher energy densities than batteries, longer runtime, instant recharging, use a liquid fuel, operate at room temperature and can rely on a passive flow of reactants, allowing to operate the cell without any additional power consumption. Therefore, in the last years, the passive DMFCs systems,

[*] Corresponding Author Email: vaniaso@fe.up.pt.
[†] Corresponding Author Email: apinto@fe.up.pt.

where the fuel and the oxidant are feed in a passive mode, have been studied and some progress towards its commercialization has already been achieved. However, the current systems still have higher costs, lower durability and power outputs due to the slow electrochemical reactions, water and methanol crossover and an inefficient products removal from the anode (gaseous carbon dioxide) and cathode sides (liquid water), which may hinder the fuel and the oxidant supply to the reaction zone.

This chapter discusses the key work done in order to improve the passive DMFC performance based on the challenges of this technology. The main goal is to provide a review on the most recent developments in passive DMFCs and on the recent work concerning the optimization of the operating and design conditions and on empirical and fundamental modelling. This chapter starts by a brief introduction recalling the passive DMFCs technology followed by a description of the fundamentals of these systems. Then an intensive review on the recent experimental and modelling works performed with these cells is presented. Towards the introduction of passive DMFCs in the market, studies regarding its lifetime and durability, as well as a cost benefit analysis where this technology is compared with the traditional ones is also presented.

Keywords: passive direct methanol fuel cell, methanol crossover, water management, two-phase flow phenomena, mathematical modelling

INTRODUCTION

In the last years, the interest in the different types of fuel cells (FCs) has increased drastically, since they were identified as a promising power source for transportation, stationary and portable electronic devices [1, 2]. This was due to their ability to convert the chemical energy of a fuel directly into electrical energy, to the fact that have zero or low emissions and the absence of moving parts. Moreover, regarding the portable applications the traditional batteries are becoming inadequate to the energy demand of the new devices due to an increase of its functionalities. Nowadays the mobile phones incorporate graphics and games, internet service, instant messaging and are even helpful to find a restaurant or a museum. This consumers demand encouraged researchers to find alternative portable power sources to overcome limitations of the conventional batteries. The target technology was the direct methanol fuel cells (DMFCs) since they offer high energy densities than batteries, longer runtime, instant recharging, use a liquid fuel, allow the operation of the cell under ambient conditions (room temperature and pressure) and with the advance of micromachining technologies, fuel cell miniaturization promises higher efficiency, performance and lower costs [3, 4].

There are two types of fuel and oxidant supply in a DMFC: an active and a passive one. Active systems use extra components to feed and manage the fuel and the oxidant, such as pump and blowers allowing the operation of the fuel cell at favourable conditions regarding the fuel cell temperature and reactants concentration and flow rate. However, these systems need an input of energy, have lower system energy densities, are more complex and have higher sizes and costs, being for that reasons unsuitable for portable applications. Passive systems use natural forces, diffusion and convection to achieve all the processes without any additional power consumption. Therefore, compared to active systems, passive ones enable a more compact and simpler design, since in this case the space assigned for pumps, fans and tubing is used for the fuel reservoir, thus increasing the whole system energy capacity. In addition, these systems do not have any additional power consumption, have a fast refuelling and the fuel can last several months. Consequently, they become more suitable for portable power sources and major efforts are being made towards the development of optimized passive DMFC systems. However, besides the common challenges of the two feed systems (higher costs, lower durability and power outputs due to the slow electrochemical reactions, water and methanol crossover and an inefficient product removal from the anode and cathode sides) the passive systems, also, suffer from the difficulty in getting a continuous and homogeneous supply of reactants.

In the last years, many advances have been achieved in the DMFC technology and different review articles, with a different emphasis, are available in literature. These include the following: anode and cathode catalysts [4-8, 12, 26, 27], MEA preparation, procedure and materials [9-13, 26, 27], cell design and performance [13, 14, 16-20, 22, 26, 27], current collectors [24, 26, 27], operating conditions [13, 14, 16, 17, 20, 26], costs

[3, 4, 5, 7, 25, 26, 27], DMFC challenges (methanol crossover, water and heat management) [4, 13, 15-18, 20, 25, 26], durability and stability [5, 25, 26, 27], applications [4, 10, 15, 17], modelling [21, 22, 23] and stacks and prototypes [17].

Based on the review articles already published and due to a lack of a recent work regarding the experimental and modelling studies and the challenges of the passive DMFCs, the main goal of this chapter is to provide a recent review concerning these issues. This chapter aims to be a useful and powerful tool for researchers who already work with this technology and the ones that want to start working in this field towards its optimization and massive commercialization.

PASSIVE DMFC OPERATING PRINCIPLE AND CHALLENGES

A DMFC is an electrochemical device that transforms the chemical energy into electric energy based on the methanol oxidation and oxygen reduction reactions and using a polymer electrolyte membrane (PEM) as the electrolyte. This polymer is permeable to protons who are the ionic charge carrier. A typical passive DMFC comprises an end plate with a fuel reservoir, a current collector, a diffusion layer (AD) and a catalyst layer (AC) at the anode side, a proton exchange membrane (PEM) and a catalyst layer (CC), a diffusion layer (CD), a current collector and an end plate opened to the surroundings at the cathode side (Figure 1). The most important component of this system is the membrane electrode assembly (MEA), formed by sandwiching the PEM between an anode and a cathode electrode (catalyst layer and diffusion layer). Upon hydration, the PEM shows good proton conductivity. At both sides of the PEM, there are the catalyst layers where the reactions take place and on each side of these layers, two diffusion layers provide the current collection and allow the distribution of the reactants towards the catalyst layers and the products removal towards the anode and cathode outlets.

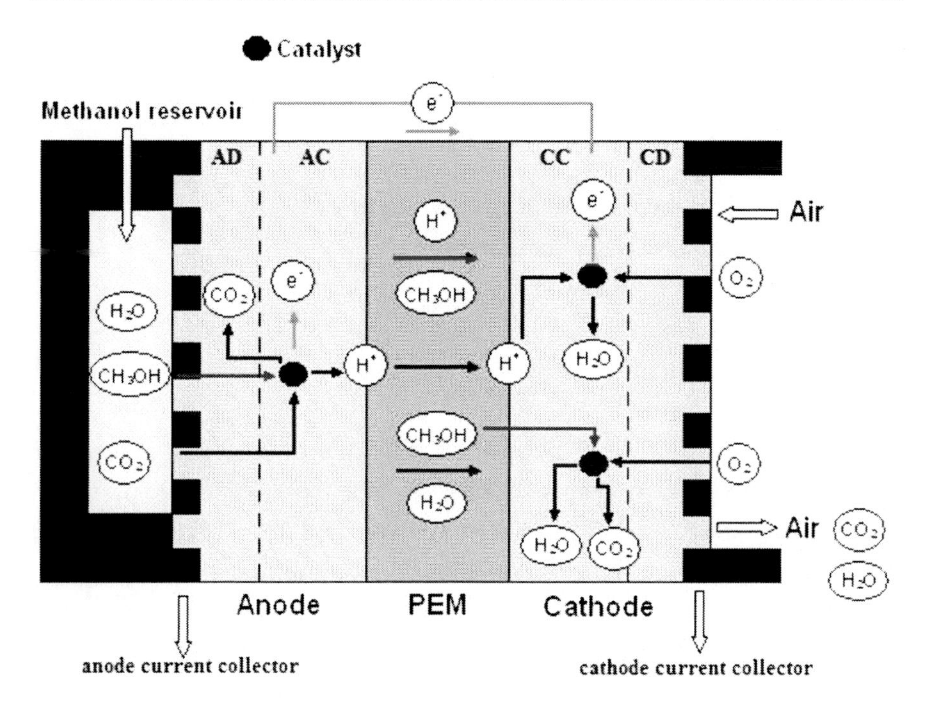

Figure 1. Schematic representation of a conventional passive DMFC.

As can be seen in Figure 1, methanol or an aqueous methanol solution is introduced on the anode reservoir. The reactant diffuses through the anode diffusion layer towards the anode catalyst layer where it is oxidized producing carbon dioxide, protons and electrons. The carbon dioxide generated emerges as bubbles and diffuses through the anode diffusion layer towards the anode end plate, since the membrane is almost impermeable to gases. The protons and electrons are transported, respectively, through the membrane and the external circuit to the cathode side. Simultaneously, air is fed to the cathode side by natural convection and the oxygen is transported by diffusion through the cathode diffusion layer towards the cathode catalyst layer, where it reacts with electrons and protons to form water. The water produced moves counter-currently towards the cathode outlet via the cathode diffusion layer and under some operating conditions, by back diffusion towards the anode.

It is commonly accepted by the scientific community that to achieve the desirable levels of energy density, costs and lifetime, the passive DMFC systems must overcome the following key challenges:

1) Low rate of methanol and oxygen kinetics;
2) Methanol and water crossover;
3) Two-phase flow patterns;
4) Higher costs;
5) Lower Durability/Lifetime.

One of the major disadvantages of using methanol as fuel is the slow anode kinetics arising from a multi-step fuel oxidation process, which result in higher anodic overpotentials. To overcome this difficulty, it would be desirable to use higher methanol concentrations, however these conditions generate higher rates of fuel crossover through the membrane towards the cathode side. This leads to a decrease of the cell performance due to a lower potential at the anode side, the formation of a mixed potential on the cathode side and a loss of the available fuel. Traditionally, a DMFC operates with diluted methanol solutions to avoid this drawback. However, more concentrated solutions are needed to achieve the energy densities requirements for portable applications. Additionally, low methanol concentrations lead to a higher water content on the anode reservoir and consequently at the anode side. This water tends to cross the membrane towards the cathode side causing a water loss from the anode, and thus, make-up of water is needed. In addition, a high rate of water crossover increases the possibility of cathode flooding, decreasing the fuel cell performance. To solve these two problems, a deep knowledge of the influence of the different design and operating parameters on water and fuel transport is mandatory for the design of ground-breaking passive DMFC systems [4, 13, 15-18, 20, 25, 26].

As, already referred, an important issue in DMFC technology is the two-phase flow pattern that occur both on the anode and cathode side due to the formation of gaseous CO_2 and liquid water, respectively. An efficient product removal is crucial to obtain higher performances, since bubbles and drops tend to remain attached to the catalyst limiting the continuous supply of reactants. Therefore, understanding the two-phase flow is crucial to develop an air breathing operation. Hence, both numerical and visualization studies are needed to clearly understand the dynamic effects of the two-phase flow on the fuel cell performance optimization [28-43].

The DMFC efficiency is limited by the reactions occurring at both electrodes (anode and cathode), since the slow kinetics of these reactions lead to significant potential losses. Moreover, even using the current state-of-the-art catalysis for each reaction (Pt/Ru at the anode and Pt at the cathode) until now, it is only possible to reach reaction efficiencies of about 25-35% [4-8, 12, 26, 27]. Additionally to achieve these efficiencies the catalyst loadings recommended are approximately 4 mg cm^{-2} at the anode and 2 mg cm^{-2} at the cathode. However, both Pt and Ru are expensive and represent a major fraction on the total system cost [4, 5, 7, 25, 26, 27]. While new low cost catalysts are not commercially available, the solution is to reduce the loadings of the available ones by improving its activity with higher surface areas, by controlling the particles size distribution, morphology and crystallinity [4-8, 12, 26, 27].

Degradation is another important challenge to overcome on the DMFC technology. Studies have shown a power density loss with the operation time of 30%. This was primarily due to the delamination of the catalyst layer and agglomeration of the catalyst particles, which lead to a loss of the electrochemical active surface area of the catalysts and membrane degradation [5, 25, 26, 27]. Although the loss of efficiency is unavoidable, the degradation rate can be minimized through an understanding of the degradation and failure mechanisms.

TWO-PHASE FLOW PHENOMENA

The two-phase flow phenomena, is one of the great challenges in DMFCs since it involves the gaseous carbon dioxide flow in the anode and the liquid water flow in the cathode, both products of the electrochemical reactions. These must be removed quickly and efficiently from the cell to allow a properly supply of the reactants (fuel and oxygen) to the catalyst layers ensuring higher electrochemical reaction rates. The water and the CO_2 can adhere to the surface of the diffusion and catalyst layers blocking the pores and preventing the fuel and the oxidant from spreading uniformly into the anode and cathode porous media. This will lead to a severe loss of performance. Hence, studies on the two-phase flow serve as a guide for the improvement of the performance of DMFC [28-43]. Flow visualization is an effective way to investigate quantitatively and qualitatively the dynamic behaviour of carbon dioxide gas bubbles in the anode channels and liquid water bubbles in the cathode of an operating DMFC. Although the number of papers published on passive DMFCs has grown, few research works have been reported in visualization experiments due to difficulty of performing such studies.

CARBON DIOXIDE IN THE ANODE

As already referred, at the anode side of a DMFC, gaseous carbon dioxide is produced by the methanol oxidation reaction. If the carbon dioxide cannot be removed efficiently from the surface of the anode diffusion layer (AD), (Figure 1) it remains covering this surface and consequently decrease the effective mass transfer area (Figure 2). Therefore, gas management on the anode side is an important and critical issue on the DMFC design. In the last years, different research groups have performed studies regarding this issue, but few are related to passive systems [28–34]. However, the CO_2 removal is very important in passive system which do not have external auxiliary equipments to control the flow of the different species.

Towards the understanding of the two-phase flow patterns on the anode side, both experimental studies, through direct visualization of the bubbles behavior [28-32], and modeling ones [33, 34] have been performed.

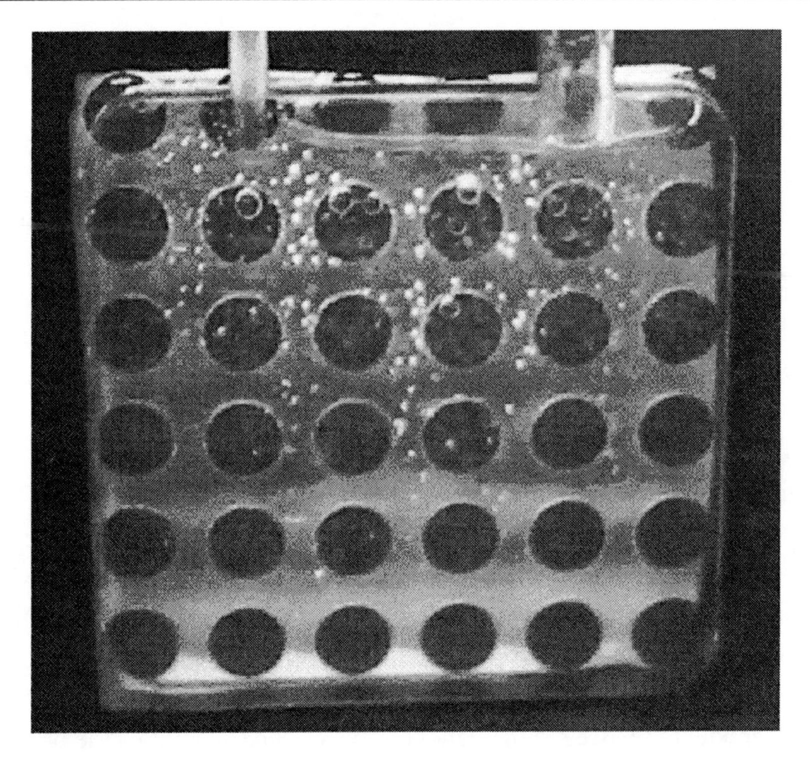

Figure 2. CO_2 bubbles behaviour on the anode of a passive DMFC.

Chen et al. [28] investigated the CO_2 microbubble removal behaviour on carbon nanotube (CNT)-supported Pt catalysts, through an in-situ visualization technique. According to the experimental data, the CO_2 microbubble removal behaviour is directly related to the high current and the nanostructure characteristics of the electrodes contributed to an enhance of the mass transfer rate. Yuan et al. [29] presented the effect of the anode and cathode structural properties on the CO_2 removal rate, as well as, the effect of the cell orientation on the performance of a passive DMFC. The results indicated that the cell performance can be enhanced using carbon cloth as anode diffusion layer and carbon paper as cathode diffusion layer. The authors recommended the use of current collectors with a circular-hole-array pattern and with a small open ratio at the anode and a parallel-fence pattern with a high open ratio at the cathode, since these layouts yield the highest performances for a wide range of methanol concentrations. Gholami et al. [30] studied the effect of the current collector design and the methanol concentration on the performance of a passive DMFC and on the CO_2 removal rate, employing two different designs and arrangements for the current collectors were tested. In the first one, non-uniform parallel channels were machined on the current collector and used on the anode side. A perforated design was used on the cathode current collector. In the second design, uniform parallel channels were machined on both anode and cathode current collectors. The results showed a higher performance for the first arrangement using a methanol concentration of 4 M (20 mW cm^{-2}) and this was explained by a higher removal rate of carbon dioxide produced by the positive slope of the channels and by a higher methanol mass transfer rate, due to a higher open ratio. Yuan et al. [31] studied the optimization of the anode design by developing a composed diffusion medium with different porosities, which included three carbon-based layers: a woven carbon-fibre fabric (WCFF), a carbon paper and a carbon-powder micro-porous layer. The intention of using the WCFF was to control the mass transfer of reactants and products at the anode, to reduce the methanol crossover and to facilitate the products removal. It was found that when using the WCFF the fuel cell achieved the highest performance at a methanol concentration of 12 M, due to a more uniform distribution of the reactant and a rapid removal of the gas bubbles. Falcão et al. [32] investigated the effect of using stainless steel meshes, as an effective way of controlling the two-phase flow phenomena, on the performance of passive micro-DMFC. The higher performances were obtained with the cells employing a metallic mesh due to an increase of the current collection, a decrease of the methanol crossover rate and a better distribution of the CO_2 bubbles. Meshes with larger open ratios produced better results, however they lead to higher compression rates and leakage problems. The best power output obtained was 29.3 mW cm^{-2} using stainless steel meshes at both anode and cathode sides.

As already referred, a better understanding of the basic transport phenomena of carbon dioxide bubbles at the anode side can be achieved combining both visualization studies and numerical simulations. Towards that, Zheng et al. [33] developed and used a two-dimensional model to investigate the multiphase flows and the interactions between the different layers on the anode side of a DMFC, focusing on the flow of a single CO_2 bubble. The simulation results show that as the power output increases, the porosity and the diffusion layer (DL) contact angle play a significant role on the size of the CO_2 bubbles, allowing bubbles to grow to their maximum size. The thickness of the DL has an effect on the CO_2 bubbles size at lower power outputs and the thickness can be estimated by the maximum size of the bubbles. The model suggests that, to achieve optimal performances, the DL in passive DMFCs should be thick enough to allow bubbles to grow to their maximum size. Liu et al. [34] proposed a new approach to promote CO_2 removal from the anode catalyst layer of a passive DMFC by introducing a Lorentz force, via the integration of magnetic field on the system. This forced convection around the CO_2 bubbles was studied through mathematical simulations. The results indicated that the magnetic field improved the maximum power density on 12.5%, due to an enhance of the CO_2 removal rate.

LIQUID WATER IN THE CATHODE

Another important aspect of the DMFCs is the possibility of water flooding at the cathode pores and structure, due to water transport through the membrane and water production by the cathode reaction. The formation of water within the cathode catalyst layer and its transport through the cathode diffusion layer represent a mass transport resistance and disturb the diffusion of oxygen towards the reaction zone, which can reduce the limiting current density of the cell. A proper water level at the cathode side is necessary to hydrate the polymer membrane, increasing in this way the proton conductivity. However, a high amount of water in the cathode side leads to water flooding at the pores decreasing the cathode performance [35] (Figure 3). An optimized water management is necessary in passive DMFC systems to overcome this difficulty thus inspiring new design concepts. In order to accurately predict the operation and design conditions that avoid flooding, visualization and modelling studies regarding the cathode side are essential to yield fundamental physics behind the flooding occurrence [35-43].

Figure 3. Cathode flooding in a passive DMFC.

Zhou et al. [35] developed an air-breathing micro (μ) DMFC with a water collecting and recycling structure to control the water distribution inside the cell. Towards that, capillary channels with a hydrophilic surface were placed along the ribs of the air-breathing cathode window, to collect the water in excess and release it into the exhaust pipe. The results showed that the power and the current density of the cell with this system were 9.7 mW cm^{-2} and 95 mA cm^{-2}, respectively, values higher than those obtained with the conventional layout 9 mW cm^{-2} and 75 mA cm^{-2}. Xu and Faghri [36] investigated the water transport characteristics in a passive DMFC considering a porous diffusion medium with both hydrophilic and hydrophobic pores. This study was based on a mathematical model that uses as input information a mixed-wet capillary pressure and a saturation relation (CPSR), determined experimentally. The liquid water transport predicted by the mixed-wet CPSR was compared with the uniform-wet Leverett CPSR. It was found that, the use of the Leverett CPSR underestimates the liquid saturation in the diffusion layer, thus overestimating the cell performance. Moreover, it was found that using the mixed-wet CPSR, the liquid saturation in the diffusion media is nearly independent of the operating current and mainly depends on the wettability of the porous structure, such as, the fraction of the hydrophilic pores. Li et al. [37] worked with a semi-passive high-concentration (HC)-DMFC, where the water management is much more critical due to the small amount of water in the fuel reservoir. They used a polytetrafluoroethylene (PTFE) layer as a methanol barrier layer between the fuel reservoir and the current collector to improve the water management and analyze the effect of different parameters on the overall HC-DMFC performance. The results showed that the methanol barrier layer could significantly decrease both methanol and water crossover and increase the fuel efficiency. The membrane resistance was the predominant factor affecting the HC-DMFC performance, especially for high current densities. Since in this case, the water back-flow towards the anode is very important to hydrate the membrane, they found that this can be enhanced increasing the cathode relative humidity and pressure. The water flooding at the cathode of a HC-DMFC with a porous PTFE plate was not relevant and a lower oxygen flow rate is desirable to decrease the water losses and to achieve better performances. Chen et al. [38] designed and fabricated a novel cathode catalyst layer with a PTFE and Nafion impregnation and with a discontinuous hydrophobicity gradient, since this distribution can be beneficial to the oxygen diffusion inside this layer and to the water removal. The authors concluded that the hydrophobic distribution could accelerate the oxygen transport in the catalyst layer and enhance the water back diffusion, improving the DMFC performance and its stability. Yousefi et al. [39] designed and studied the effect of the cell orientation and the environmental conditions on a passive DMFC and found that the best performance was achieved under vertical orientation since with the horizontal one (anode downward) occurred cathode flooding and CO_2 accumulation on the anode side. According to the results, a higher environmental temperature leads to an improved of the electrochemical kinetics, a decrease of the cell internal resistance and a decrease of the cathode flooding, resulting in better performances. However, an increase on the air relative humidity lead to a decrease of the performance. Wang et al. [40] used the plasma electrolytic oxidation (PEO) to prepare a super-hydrophilic coating on the surface of an aluminium based cathode current collector in order to solve the water flooding problems in a micro passive DMFC. The results showed that the liquid water spreads quickly as a plane on the surface of the porous PEO proving that this treatment can prevent the cathode flooding. Therefore, the fabricated PEO-cell showed better power outputs than the normal-cell and an extremely higher stability. Following this research line, Zhang et al. [41] developed a novel water management system for a metal-based micro passive DMFC that includes a hydrophilic coating, through the plasma electrolytic oxidation (PEO) technique, in some channels of the cathode end plate, a stainless steel felt as cathode current collector and an aluminium-based cathode end plate with a perforated flow field (holes). Due to the highly hydrophilic nature of the PEO, the results showed that the channels efficiently collect/remove the liquid water produced on the cathode reaction suppressing the water accumulation along the air-breathing holes, preventing the water flooding and improving the cell performance and stability. It was also, found that the cathode end plate design allows the removal of the liquid water to the outside of the cell and enable the water recycle.

Bahrami and Faghri [42] investigated the fuel delivery and its relation with the water and methanol crossover using a two-dimensional, two-phase flow, steady state, non-isothermal, non-equilibrium and multi-fluid model for a passive DMFC. Is this work, the authors used a porous layer, called the fuel delivery layer, between the anode diffusion layer and the anode reservoir in order to use a high methanol concentration. A hydrophobic microporous layer was used at the cathode to decrease the methanol and the water transport through the membrane. The results showed that an increase of the fuel delivery layer thickness was not effective to control the methanol transport through the membrane without controlling simultaneously the water crossover rate. On the contrary, the use of a cathode microporous layer, leads to a reduction of water and methanol crossover and significantly reduces the liquid saturation at the cathode diffusion layer reducing water flooding.

In spite of these studies, much remains to be understood on the fundamental processes of two-phase flow phenomena that affect the anode and cathode sides of a passive DMFC and its relation with the different materials used in the cell layers. As can be seen, the introduction of metallic meshes between the membrane electrode assembly and the current collectors, using current collectors with different designs and different orientations for the fuel cell is an effective way to control the carbon dioxide flow and release at the anode. Regarding the water on the cathode side, the use of additional layers and the common ones with hydrophobic/hydrophilic treatments is an effective way to reduce the cathode flooding through a more efficient water removal from the cathode outlet and a higher water back flow rate towards the anode side.

MASS TRANSPORT PHENOMENA

A deep understanding of the mass transport phenomena is essential to improve the fuel cell performance and stability. Many authors propose different solutions to eliminate or reduce the fuel loss across the membrane, as well as, to solve the water management issues through different approaches: modeling/simulation and experimental studies [43–64]. Xiao et al. [43] developed a two-dimensional, non-isothermal, transient, multi-phase model to study the heat and mass transport in a passive and semi-passive liquid-feed DMFC. The results showed that, as expected, an improved mass transport system leads to an improved fuel cell performance. However, to achieve even higher performances an improved heat transport system is, also, needed. Xu et al. [44] investigated the characteristics of the methanol and water crossover in a passive DMFC and the results demonstrated that the cell performance and the methanol and water crossover through the membrane strongly depended on the heat and water management issues. With an increase on the current density, the methanol crossover increased, while the fuel efficiency first increased and then slightly decreased. The water crossover increased almost linearly with the current density, but was almost the same for the different methanol concentrations tested. The results, also, showed that although the water in the cathode leads to the cathode flooding and increases the oxygen mass transport resistance, it also, allowed to reduce both water and methanol crossover rates, thus increasing the fuel efficiency. The authors found that a reduction of water crossover rate was always followed by a reduction of the methanol crossover rate.

METHANOL CROSSOVER

In DMFCs, methanol crossover occurs due to the inability of the Nafion membranes to prevent methanol, which has a small size, from permeating its polymer structure. Diffusion and electro-osmotic drag are the prime driving forces for its transport through the membrane. The methanol that reaches the cathode side reacts in the platinum catalyst sites on the cathode, leading to a mixed potential, which causes a decrease on the cell voltage. Methanol reaching the cathode also lead to a decrease of the fuel efficiency, due to a fuel loss to the cathode side, lowering the energy density of the system. In order to improve the performance of a DMFC, it is necessary to eliminate or reduce the loss of fuel across the membrane. As already referred, the major challenge is to find an optimized cell design that allows working with high methanol concentrations or even pure methanol without generating higher rates of methanol crossover. In this section, special attention is given in showing the most recent studies performed on this issue and some possible solutions to solve this problem.

Feng et al. [45] studied a passive DMFC fed with pure methanol and with an active area of 9 cm^2 using a pervaporation membrane (PM) to control the methanol transport. The results demonstrated that the proposed configuration can successfully operate with pure methanol allowing to obtain a maximum power density of 21 mW cm^{-2}. The results showed that at a current of 100 mA, the energy density of the cell supplied with pure methanol increased about 6 times than that of a fuel cell fed with a dilute methanol solution. Park et al. [46] designed a MEA with a multi-layer electrode, which includes a fuel control layer to decrease the fuel crossover, a hydrophobic layer to reduce water crossover and a hydrophilic layer to improve the water back flow towards the anode, which was used on the anode side of a passive DMFC operated with high methanol concentrations. The results showed that the MEA proposed improve the cell performance by decreasing both methanol and water crossover rates. Yuan et al. [47, 48] designed and studied the performance of passive DMFC using a porous metal-fiber sintered felt (PMFSF) on the anode side of the cell, as a mass-transfer-controlling medium, to mitigate the negative effects of the methanol crossover. The experimental results revealed a significant enhancement on the cell performance, using a PMFSP with a thickness of 2 mm, especially at higher methanol concentrations. Years later, the same authors prepared and added graphene-carbon nanotubes (G-CNTs) into the anodic micro porous layer (MPL) to form a crack-free anode diffusion layer for a passive DMFC. The intention was to decrease the methanol crossover rate and improve the electron contact between the diffusion and the catalyst layers [49]. In this work, a maximum power density of 41.6 mW cm^{-2} was achieved at an operating temperature of 25 °C, due to an increase of 36.1% of the electrochemical active surface area (estimated through carbon monoxide stripping measurements), a decrease on the charge transfer and contact resistance and a decrease on the methanol crossover rate. Huang et al. [50] developed a new porous anode electrode by adding magnesium oxide (MgO) nanoparticles as pore-former into the catalyst layer and MPL. The novel MEA showed improved performances and a striking reduction on the use of noble metals as catalysts, due to an increased electrochemical surface area and a decreased anodic charge-transfer resistance. Abdelkareem et al. [51] investigated the effect of using carbon nanofibers (CNFs) as a 1D nanostructure on the cathode MPL of a passive DMFCs. The polarization curves and the in situ cyclic voltammetry measurements indicated that the CNFs content has a significant influence on both mass transport and electrochemical surface area. The highest catalyst utilization was observed with a CNFs content of 40 wt.%, however the maximum power density (36 mW cm^{-2}) was achieved at a CNFs content of 20 wt.% due to a lower mass transfer resistance under these conditions. Wu et al. [52] studied the effect of using polypyrrole nanowire networks (PPNNs), as a MPL on the carbon paper surface, on the fuel cell performance. The results demonstrated that the use of PPNNs can effectively increase the surface area of the carbon fiber improving the catalyst utilization, promoting the methanol mass transport and consequently decreasing the methanol crossover. Zhang et al. [53] developed a novel anode mass transfer barrier layer using reduced graphene oxide (rGO) deposited on a stainless-steel fiber felt (SSFF). The SSFF-rGO layer was used as both anode current collector and methanol mass transfer barrier. The results showed that the SSFF-rGO layer can increase the methanol mass transport resistance, decreasing the methanol crossover rate, but allowing a proper methanol concentration on the reaction zone, thus leading to better performances. Oliveira et al. [54] investigated the effect of the anode diffusion layer properties on the performance of a passive DMFC aiming to achieve optimized performances working the cell at high methanol concentrations. The results showed that using carbon paper with lower thicknesses and with a MPL lead to higher fuel cell performances due to a decrease of the anode overpotential and a higher methanol oxidation rate on the catalyst layer. However, the best power density, 24.3 mW cm^{-2}, was achieved using carbon cloth with higher thicknesses as anode diffusion layer and a methanol concentration of 3 M mainly due to the lower methanol crossover rates generated.

WATER MANAGEMENT

In order to compete with traditional batteries, the most important requirement for a passive DMFC is to have higher energy densities. Recent studies [55-64] indicate that the water management is a critical challenge for DMFCs achieve the desirable energy levels. The amount and distribution of water within the fuel cell strongly affects its efficiency and reliability. As was described in the last section, another important challenge to overcome in DMFC employing Nafion membranes is the methanol crossover and one solution to solve this problem is to use lower methanol concentrations. This leads to the presence of higher amounts of water at the anode side and consequently a large amount of water needs to be carried out in the system, thereby reducing the energy content of the fuel mixture. The presence of a large amount of water floods the cathode and reduces its performance. Therefore, an important engineering issue is to remove water from the cathode to avoid severe flooding and subsequently supply water to the anode to make up water loss due to water crossover through the membrane. Low water flux through the membrane is desirable for DMFCs, as the anode does not require an excessive amount of water replenishment and the cathode is less susceptible to severe flooding. Formally, the water flux through the membrane, caused by diffusion and electroosmotic, can be quantified in terms of net a water transfer coefficient (α-alfa value) [62]. The ideal value for this coefficient is a negative one, meaning that no water is necessary from the anode fuel solution and the water needed to oxidize methanol comes from the water produced at the cathode side.

Shaffer and Wang [55] developed and used a 1D, two-phase model that accounts for the capillary-induced liquid flow in porous media and show how a hydrophobic anode MPL controls the water crossover rate. The results indicated that a lower water crossover rate could lead to a lower methanol crossover rate due to a dilution effect of methanol in the anode catalyst layer. Finally, through a parametric study it was possible to show that a thicker anode MPL with a higher hydrophobicity and lower permeability can be an effective solution to control the water crossover. Cao et al. [56] proposed a novel MEA design with a double MPL on the cathode side, employing a Ketjen Black (KB) carbon as an inner-MPL and Vulcan XC-72R (XC-72R) carbon as an outer-MPL. The experimental results showed that this structure provided higher oxygen transfer rates and a reduction of the water crossover rate, leading to higher cell performances and stability. Shaffer and Wang [57] used the low-α MEA concept, with a lower or negative water crossover using a hydrophobic anode and cathode MPL, and an anode transport barrier (TB) between the backing layer and the hydrophobic MPL. To show the effectiveness of this novel design the authors used a 1D, two-phase, DMFC model. According to the results, a thicker, more hydrophilic and permeable anode transport barrier and a thicker, more hydrophobic and less permeable anode MPL lead to lower methanol and water crossover rates. Wu et al. [58] analyzed the water retention rate in the membrane of a DMFC operating with pure methanol, through the use of a thin layer (retention layer) consisting of nanosized SiO_2 particles and a Nafion ionomer coated onto each side of the membrane. The hygroscopic nature of the SiO_2 allows maintaining a higher water concentration level at the cathode side enhancing the water transport towards the anode, while the anode retention layer can retain the recovered water from the cathode side and ensure a proper water concentration at the anode catalyst layer, improving the anode reaction kinetic and the overall cell performance. Peng et al. [59] proposed and demonstrated cathodic water/air management device (WAMD) for a passive micro-direct methanol fuel cells. The apparatus consists of micro-channels and air-breathing windows, with a hydrophobicity treatment on the rib structure as a way for a rapid water removal and to maintain a high gas pathway for air convection. The WAMD resulted in a water removal rate of 5.1 $\mu l\ s^{-1}\ cm^{-2}$, which is about 20 times faster than the water generation rate in a DMFC operated at 400 $mA\ cm^{-2}$. The water vapors were, also, collected and the water was recirculated by the micro-channels, which act as a passive water recycling system. Wu et al. [60] showed the effect of the cathode diffusion layer design, namely the PTFE content in the backing layer (BL) and MPL and the carbon loading in the MPL, on the water and oxygen flux and on the performance of a passive DMFC operating with neat methanol. The results indicated that the cell performance increased with the PTFE content both in the cathode BL and MPL due to a reduction of the water crossover. Zhang et al. [61] studied the performance of passive air-breathing DMFC fed with pure methanol with a homemade wet-proofing carbon paper used as backing layer on the cathode side. Different amounts of a hydrophilic ionomer (5 wt % Nafion solution) and a hydrophobic additive (PTFE) were added to achieve a hydrophobic gradient for dynamic water back-flow. The results demonstrated that the energy density of the DMFC fed with pure methanol using a backing layer with 40 wt % PTFE on the cathode was increased by six times when compared to a conventional DMFC fed with a methanol concentration of 2 M. The results confirmed that a highly hydrophobic cathode diffusion layer exhibited an excellent water management and was crucial for water back-flow during the long-term operation of the passive DMFC. Oliveira et al. [62] studied the water transport through the membrane towards the cathode using a mathematical model. The authors analyzed the effect of methanol concentration, membrane thicknesses, diffusion layer materials and thicknesses and catalyst loading on the water crossover rate and consequently on the passive DMFC performance. The results showed that the methanol concentration clearly affects the water transport rate through the membrane, since lower methanol concentrations resulted in higher water concentration gradients between the anode and cathode side and consequently higher water crossover rates. Concerning the different materials for the diffusion layers tested, better performances were achieved using carbon cloth, mainly due to its higher thickness, which lead to lower methanol and water crossover rates. Lower water crossover rates were, also, achieved with higher catalyst loadings on the anode side, since higher loadings lead to thicker catalyst layers and higher mass transport resistances. Regarding the membrane thickness, it was found that thinner membranes exhibit lower water crossover rates. Deng et al. [63] proposed the use of CNT paper as cathode diffusion layer for a passive μ-DMFC, to control de water management issues. The passive μ-DMFC using the CNT paper showed higher performances than that with carbon paper and allow the operation with high methanol concentrations due to a higher water back diffusion and lower methanol crossover rates. Yan et al. [64] investigated a new MPL with a hydrophilic-hydrophobic dual-layer to trap and retain the water and simultaneously create a capillary pressure to prevent the lack of water in the anode side of a passive DMFC operating with neat methanol. The results showed a higher water concentration on the anode side, which improved the anode reaction kinetics and the overall cell performance.

Based on the work done regarding the methanol and water crossover phenomena, it can be concluded that the methanol and water crossover is unavoidable. Moreover, both phenomena are interconnected and the reduction of one can lead to the reduction of the other. Based on the methanol crossover state-of-the-art it can be seen that its rate can be mitigated/minimized through the improvement of the activity of the methanol electroxidation catalysts, the use of different catalyst loadings, different materials for the anode diffusion layer and with different thicknesses and methanol mass transport barriers/layers. Regarding the water management issue, a low or even negative net water transport coefficient is mandatory to achieve higher power outputs. This can be accomplished using different layers or mass control layers of different materials, with different properties and or treatments (hydrophobic or hydrophilic) and thicknesses both in the anode and cathode sides.

SINGLE CELL PERFORMANCE

A passive DMFC is a multiphase system involving simultaneous mass, charge and energy transfer. All these processes are intimately coupled, resulting in a need for searching the optimal cell design (current collectors design, membrane, diffusion and catalyst layer properties, fuel cell layout) and operating conditions (cell temperature, methanol concentration). A good understanding of these complex phenomena is thus essential to optimize the cell performance and this can be achieved through a combined mathematical modelling and detailed experimental approach. In fact, some experimental work has been done in order to achieve optimal performances [65-125]. To improve the power outputs of a passive DMFCs there are an increased interest in reduce the mass transport limitations and the kinetic and ohmic limitations. Towards that, some work has been done in order to evaluate the effect of the methanol concentration (Met C) on the cell performance and to improve the design of the current collectors (CC), the catalyst loading (Cat L), the characteristics of the diffusion layer (DL) and membrane (Mem) in terms of materials and thicknesses (Table 1).

Table 1. Single cell performance works based on the parameters studied and the maximum power output achieved

Ref	Details	Parameters studied					Max. Power (mW cm^{-2})
		Met C	DL	CC	Cat L	Mem	
[65]	Developed a microfluidic-structure anode flow field.	X					23.5
[66]	Designed a new structure with two methanol reservoirs separated by a porous medium layer.	X					16.4
[67]	Analyzed operational characteristics, with focus in the fuel concentration, of a small-scale passive DMFC.	X	X			X	5.97
[68]	Developed a passive DMFC with 100 cm^2.	X		X			5.2
[69]	Developed two stacks, one with dilute methanol solutions and the other with pure methanol	X					30.1
[70]	Investigated the use of a porous metal fiber as an anodic methanol barrier.	X					–
[71]	Developed a passive and self-adaptive DMFC using a polypropylene based pervaporation film to supply the methanol in vapor form.	X					42
[72]	Developed a micro-porous anode current collector.	X		X			41.4

Table 1. (Continued)

Ref	Details	Parameters studied					Max. Power (mW cm^{-2})
		Met C	DL	CC	Cat L	Mem	
[73]	Developed an integrated anode structure based on a microporous titanium plate (Ti-IAS), as current collector and with a pervaporation film.	X		X		X	40
[74]	Investigated the effect of the key system parameters that influence the performance of passive DMFC operated with concentrated fuel.	X	X				24
[75]	Investigated the effects of a porous metal fiber used as a methanol barrier on the anode side.	X		X			–
[76]	The effect of cell orientation on cell performance employing a porous carbon plate.	X					35
[77]	Studied some operating parameters on the performance of a passive DMFC.	X					–
[78]	Improved a tubular-shaped passive DMFC to achieve stable operation with high concentrations of fuel.	X				X	16.5

Ref	Details	Parameters studied					Max. Power (mW cm^{-2})
		Met C	DL	CC	Cat L	Mem	
[79]	Proposed a dual-chamber anode structure, with different methanol concentration, for a passive μ-DMFC.	X					–
[80]	Investigated the optimized design of a metallic passive μ-DMFC for portable applications.	X	X				19.2
[81]	Developed a bi-layer composite membrane, of sulfonated graphene oxide (SGO)/Nafion and sulfonated activated carbon (SAC)/Nafion.	X				X	29.1
[82]	Evaluated the performance of Nafion recast with addition of palladium-silica nanofibres (Pd-SiO_2).	X				X	10.4
[83]	Investigated a methanol-blocking membrane prepared by assembly of poly(diallyldimethylammonium chloride) (PDDA) and graphene oxide (GO) nanosheets onto the surface of Nafion® membrane.	X				X	20.6

Table 1. (Continued)

Ref	Details	Parameters studied					Max. Power (mW cm^{-2})
		Met C	DL	CC	Cat L	Mem	
[84]	Prepared and characterized a series of sulfonated poly (p-phenylene-co-aryl ether ketone)s (SPP-co-PAEK) membranes.					X	24.5
[85]	Developed proton-conducting membranes based on a series of side-chain-type sulfonated poly(ether ketone/ether benzimidazole)s via benzimidazolization and acylation.					X	39.3
[86]	Synthesized and evaluated a series of sulfonated cardo poly(arylene ether sulfone)s (SPES-x) membranes based on a novel bisphenol monomer containing electron rich tetraphenylmethane groups.					X	29.4
[87]	Developed a novel membrane based on a new series of cardo poly(arylene ether sulfone/nitrite)s FSPES-x with pefluoroalkyl sulfonic acid groups.					X	25.9

Ref	Details	Parameters studied					Max. Power (mW cm^{-2})
		Met C	DL	CC	Cat L	Mem	
[88]	Analyzed the effect of the Nafion®115 membrane on the cell performance considering the its roughness factor.				X	X	27.2
[89]	Investigated the effect of the decrease of Nafion ionomer size in the anode catalyst layer on the cell performance.				X		32.9
[90]	Prepared and characterized a series of Nafion composite with microporous organic polymer networks.					X	21.5
[91]	Polymerized a PEM to high the proton conductivity by plasma technique for a semi-passive μ-DMFC					X	20.1
[92]	Sinthesized and investigated a tetra-sulfonated poly(*p*-phenylene-*co*-aryl ether ketone) membranes with microblock moieties.					X	31.9
[93]	Designed a Nafion sandwich, with a monolayer graphene between two membranes, to improve the permeability of protons.					X	25.2

Table 1. (Continued)

Ref	Details	Parameters studied					Max. Power (mW cm^{-2})
		Met C	DL	CC	Cat L	Mem	
[94]	Studied the catalytic activity of Pd and of PdxCo alloy nanoparticles for the oxygen reduction reaction.				X	X	–
[95]	Synthesize Pd-based alloy electrocatalysts and evaluate it as cathode catalyst in a passive DMFC.				X	X	2.53
[96]	Developed, with the aid of electrospinning, nanofiber 3D network structure to use as anode catalyst layer.				X		43
[97]	Investigated the effect of double-layered cathodes with a reduction up to 50% on the catalyst loading, on cell performance.				X		39.4
[98]	Designed a membrane with synthesized electrocatalysts.	X	X		X		11.9
[99]	Developed a methanol-tolerant catalyst, with PtM/C and PtMRu/C combinations (M = Co or Fe) to study the enhanced for the oxygen reduction reaction.				X		–

Ref	Details	Parameters studied					Max. Power (mW cm^{-2})
		Met C	DL	CC	Cat L	Mem	
[100]	Prepared the nano-network structure by zinc oxide nanorods within anode catalyst layer and MPL to reduce the noble metal catalyst loading.				X		40.2
[101]	Investigated the methanol oxidation on Pt-Sn electrocatalyst supported on carbon-polyaniline composite.				X		4
[102]	Prepared the ordered Pt nanorod array onto MPL by electrodeposition to use as the catalytic cathode.		X		X		17.3
[103]	Synthetize a binary carbon supported Pd3CO nanocatalyst for oxygen reduction reaction.				X		3.81
[104]	Analyzed three anode electrodes made with different carbon materials.	X	X				23
[105]	Investigated the degradation mechanisms in a membrane, with CNF as anode porous layer, under different modes of operation.		X				22.9

Table 1. (Continued)

Ref	Details	Parameters studied					Max. Power (mW cm^{-2})
		Met C	DL	CC	Cat L	Mem	
[106]	Analyzed the effect of diffusion layer compression on the performance of a passive DMFC.	X			X		5.78
[107]	Developed a cathode diffusion layer with mesoporous carbon.		X				31.4
[108]	Investigated different structural factors on the cell performance by means of orthogonal array analysis.	X	X	X		X	10.7
[109]	Studied the effect of different types of carbon materials (graphene nano-sheets) as anode diffusion layers.		X				13.7
[110]	Analyzed the feasibility of using a stainless steel wire mesh as current collector in a passive DMFC.			X			2.26
[111]	Investigated the performance of a passive DMFC with different structural configurations.	X		X			–
[112]	Studied structural aspects on the cell performance.		X	X		X	10.7

Ref	Details	Parameters studied					Max. Power (mW cm^{-2})
		Met C	DL	CC	Cat L	Mem	
[113]	Investigated the effect of anode and cathode flow field geometry on passive DMFC	X		X			–
[114]	Analyzed the clamping effects on passive DMFC using electrochemical impedance spectroscopy.	X					–
[115]	Studied the effect of size and shape of active area on the performance of passive DMFC.	X					3.5
[116]	Proposed a new structure to lower power applications, with a metal mesh directly welded onto the membrane.				X	X	1.62
[117]	Developed a tubular-shaped passive DMFC with a higher instantaneous volumetric power energy density.	X				X	24.5
[118]	Developed a stack (8 unit cells) with a porous carbon plate.	X					30/stack
[119]	Developed a six dual cells connected in series for portable electronic devices.	X	X			X	25/stack
[120]	Evaluated practical operating performances with a small passive DMFC stack.	X					26/stack

Table 1. (Continued)

Ref	Details	Parameters studied					Max. Power (mW cm^{-2})
		Met C	DL	CC	Cat L	Mem	
[121]	Investigated different membranes to be used in stacks.	X	X		X		20/stack
[122]	Designed and fabricated a “4-cell” modular passive DMFC stack and tested in electronic devices.	X					27/stack
[123]	Investigated the effects of a lightweight current collector based on printed-circuit-board technology on a single and 8-cell mono polar passive DMFC			X			18.3/stack
[124]	Developed a simple and functional DMFC stacks designs with the aim of reducing costs.	X					30/stack
[125]	Designed and tested a 6-cell stack passive DMFC for long-term operation, more than 3000 h.	X					21.5/stack

The effect of methanol concentration on the performance of a passive DMFC generally reflects two different phenomena. An increase on methanol concentration lead to an increase of the coverage of the electrocatalyst sites by methanolic species, but also increase the concentration gradient between the anode and cathode side with a consequent increase of the methanol crossover rate through the membrane. At the cathode side, methanol reacts with the oxygen to form a mixed potential, causing thereby a lower cell performance. This requires a delicate balance among the positive effects on the methanol oxidation kinetics and negative ones on methanol crossover in order to enhance the cell performance. Another point that should be accounted for is the fact that the polarization behaviour in the mass transfer region is directly related to the methanol concentration, so an increase in the limiting current density is achieved with an increase of the methanol concentration. It should, also, be pointed out, that in passive feed systems, the fuel pump is eliminated and therefore the fuel is supplied to the anode from a fuel reservoir by natural forces such as natural convection and diffusion. This simple design causes lower system performances due to the difficulty of getting a continuous and homogeneous supply of the fuel to the anode, so a more concentrated fuel solution is desirable. Among the different experimental studies reported in literature, most of them, highlight the need to work with more concentrated methanol solutions to achieve the desirable levels of power outputs and use different structural parameters to act as a methanol barrier layer towards the reduction of the crossover rate [65–80].

The polymer electrolyte membrane (PEM) being an important component of a fuel cell must exhibit a high proton conductivity, be a barrier to the contact and mixing of fuel and oxidant and must be chemically and mechanically stable in the fuel cell environment. The most commonly used membranes for DMFCs are Nafion® membranes (made of a perfluorocarbonsulfonic acid ionomer, a copolymer of tetrafluorethylene and perfluorosulfonate monomers). However, since these membranes suffer from high rates of fuel crossover different approaches are being adopted towards the development of membranes from alternative materials that meet the necessary requirements: high proton conductivity, low methanol permeability, high chemical and mechanical stability and low cost [81–93].

The microstructure of the catalyst layer is very important for the electrochemical reactions and for the diffusion of the involved species. Therefore, these layers must be porous to allow the access of the reactants into the reaction sites and the removal of the products formed in the electrochemical reactions. Moreover, the DMFC efficiency is limited by the reactions occurring at both electrodes (anode and cathode), since the slow kinetics of these reactions lead to significant potential losses. Even using the current state-of-the-art catalysts for each reaction (Pt/Ru at the anode and Pt at the cathode) until now, it is only possible to reach reaction efficiencies of about 25-35%. Additionally, to achieve these efficiencies the catalyst loadings recommended are approximately 4 mg cm^{-2} at the anode and 2 mg cm^{-2} at the cathode side. However, both Pt and Ru are expensive and represent a major fraction on the total system costs. Therefore, to decrease the catalyst costs, some studies have been done in order to evaluate the influence of the catalyst loading on the cell performance and other ones on the development of new catalyst, low cost ones, using low cost materials [94–103]. It should be noted that there are two essential properties of the electrode that may be affected when changing the catalyst loading: electronic conductivity and electrode thickness. At low current densities, the activation overvoltage is a major portion in the total overvoltage at the anode, so an increase on the Pt/Ru loading reduces the activation overvoltage at the anode, increasing the cell performance. With the increasing of the loading, the thickness of the catalyst layer increases and therefore the mass transfer resistance through this layer becomes greater. In spite of this, the cell performance increases with the metal loading because a thicker anode catalyst layer creates a higher resistance to methanol transport thereby controlling the rate of methanol reaching the membrane and reducing the methanol crossover. However, at high current densities, a higher loading leads to a decrease of the cell performance due to a higher concentration overpotential loss caused by the higher mass transfer resistance rate of methanol through this layer. On the cathode side a reduction on the catalyst loading leads to a decrease in the cell performance due to a reduction of the active surface area, an increase on resistivity and consequently a decrease in the electronic conductivity. An increase on the catalyst loading lead to a thicker layer, which will lead to a higher mass transport resistance. However, this may be advantageous at the cathode, since the mixed potential formation may be avoided in some extent. In a thicker electrode, not all the catalyst particles may be reached by the permeated methanol flux, so more active sites are free for oxygen reduction reaction.

A possible solution to work with lower catalyst loadings and ensure the electronic conductivity of the electrode it to use carbon supported materials. However, the use of carbon supported catalyst with a much lower bulk density is associated with a higher thickness of the active layer which is an important parameter for the cell performance. On the anode side a thicker electrode may lead to a higher mass transport resistance of methanol leading to a decrease of the fuel cell performance. On the other hand, as explained above, this higher resistance may be beneficial on the cathode side. Therefore, the use of carbon supported catalysts and their optimization in the electrode structure has the potential to significantly reduce the metal loading contributing to a reduction of the overall costs of passive DMFC systems.

Besides providing a good distribution of reactants and products, the diffusion layers must electrically connect the catalyst layer to the current collector, allow heat removal and provide mechanical support to the membrane, preventing it from sagging into the current collector channels/holes. The operation of a DMFC requires that methanol has good access to the anode catalyst layer while the carbon dioxide move away freely from the catalyst sites on the catalyst surface. Ideally, these flows should be isolated such that discrete paths for gas and liquid exist, rather than a two-phase flow patter, with gas bubbles moving against a liquid flow. The simplest way to approach this ideal situation, is to make the diffusion layer surface hydrophobic treating it with Teflon (PTFE-treated). The cathode of the DMFC may be similarly affected by possible problems of flooding, but in comparison to the anode this is a less critical issue. Some work has been done in order to explore the effect of the Teflon (PTFE) content, as an effective way to mitigate the two-phase flow problems, and obtain optimized performances [104-109].

In passive DMFCs, the electrons recovery and the fuel and oxidant supply to the reaction zone is ensured by the current collectors. Therefore, the current collectors must have high electrical conductivity, low thermal conductivity, high corrosion resistance and low cost. These may be in a form of channels, of a specific patter, covering the entire area or may have a porous structure, with different open ratios and sizes. Several studies were performed in order to study the effect of the current collector material and design on the fuel cell performance [108, 110–113].

Besides the optimization of the different fuel cell components, it is also possible to enhance the performance of a passive DMFC through the clamping effects [114], shape and size of the active area [115], different cell layouts [116, 117]. In addition, it is possible to increase the power output, for specific applications (volume and weight), by connecting single cells to obtain a stack configuration [118–125]. A fuel cell stack should have a uniform distribution of reactants in each cell, a proper heat management, lower ohmic losses and no leakage of reactants. To achieve these requirements some works have been done towards the development of high performance and efficient passive DMFC stacks [118–125].

The power output of passive DMFCs have been considerably increased in the last years, but it must be further increased in order to achieve the levels needed for portable applications. Towards a cost reduction while new low cost catalysts are not commercially available, the solution is to reduce the loadings of the available ones by developing more active catalyst, with higher surface areas, by controlling the particles size distribution, morphology and crystallinity. The development of a membrane electrode assembly by changing the catalyst layer loading, diffusion layer properties and membrane thickness, with a low methanol and water crossover rate and cost, will be the key issue for develop a reasonable DMFC system.

MATHEMATICAL MODELLING

Fuel cell modeling has received much attention over the past decade in an attempt to better understand the different phenomena occurring within the cell. Therefore, mathematical models have been used as a fundamental tool for the design and optimization of the fuel cells systems.

The development of a DMFC involves multidisciplinary knowledge on electrochemistry, material science and engineering. Moreover, the performance of these devices is affected by different phenomena and parameters. Investigation of the effect of each one independently and/or possible combinations of them using experimental work is costly and time consuming, so the development of mathematical models is a crucial tool to define the more suitable set of parameters to use in the design and optimization of DMFC systems with reduced time and money. Different modeling approaches are available in literature and these include the basic models (analytical and semi-empirical models) and the phenomenological ones (mechanistic models) [126–157]. The basic models, usually one-dimensional (1D) models, are an adequate tool for understanding the elementary effect of the different operating conditions and design parameters on the cell performance, water and methanol crossover and can be a useful tool for rapid calculations. However, they neglect some of the most important features such as transient performance and the spatial non-uniformities. The semi-empirical models combine theoretically differential and algebraic equations with empirical determined correlations. The great advantage of these models is their simple structure and the small computational effort to perform calculations. However, the estimated parameters from the experimental data are normally specific to a certain type of cell and valid for a limited range of operating conditions. These models are very useful to perform quick predictions for existing designs but fail to predict innovative ones. Mechanistic models are transport models using differential and algebraic equations whose derivation is based on the electrochemistry and physics governing the phenomena taking place in the cell. These equations are numerically solved by different computational methods, involving extensive calculations, but accurately predict all the phenomena occurring in the cell. Moreover, since the transport phenomena in a DMFC is inherently three-dimensional due to the two-phase flow pattern both at the anode and cathode, a deep understanding of the cell operation is needed and CFD-based models can provide very important information regarding the two-phase flow effects on the cell performance [139, 148, 151, 156, 157]. These models enable a user-defined multidimensional cell geometry and a sophisticated treatment of the transport phenomena based on the mass, momentum, energy and charge conservation equations, thus providing great insights towards the understanding and optimization of the cell design. However, the high computational time demands make its practical use very low. Therefore, these models are only suitable when focusing on the design of passive DMFCs, where real-time system-level simulations (including DMFC stacks) are considered unworkable.

The majority of the fuel cell models aims to describe the interactions occurring among the several physical and electrochemical phenomena within the different layers of the cell [127-131, 134, 137, 138, 141, 145, 149-152, 154]. However some of them allow to study in detail the effect of a specific parameter on the cell performance, such as methanol concentration, cell design or diffusion layer properties [66, 132, 142, 143, 146-148, 150-152] and/or phenomena, such as, methanol and water crossover [133, 139, 140, 144, 153-155]. Since thermal management is a key issue in the portable DMFC systems it is important to develop models accounting for this effect, but until now few passive DMFC models account this effect [126, 130, 134, 146].

In spite of the modelling work on passive DMFCs developed in the past years, a number of unsolved issues demand for intensive research. One of the most important areas to investigate is the numerical modelling of two-phase flows (both in the anode and cathode sides) and parallel experimental studies on visualization of these phenomena.

In order to move one step forward in the passive DMFC modelling it is desirable to use the advantages of each modelling approach (simpler and more complex models) coupling 1D analytical models describing the heat, mass and electrochemical phenomena occurring in the membrane electrode assembly with more complex models describing the two-phase flow patterns in the other layers. The final goal of this innovative approach is to obtain simpler 1D+2D/3D models able to correctly describe the cell performance, as well as, the two-phase flow patterns and their effects on the cell performance and methanol and water crossover rates, with reduced simulation times.

LIFETIME/DURABILITY

The lifetime of a passive DMFC can be described as the ability to maintain a constant power output over the time. Studies on this issue have shown a power density loss with the operation time of 30% and this was primarily due to permanent losses, such as delamination of the catalyst layer and membrane and agglomeration of the catalyst particles, which lead to a loss of the electrochemical active surface area of the catalysts [25, 26, 158-168]. However, in order to passive DMFCs become an accepted alternative power source for portable applications, the target requirements from the DOE (U.S. Department of Energy) are 5000 hours with an efficiency loss until 20%. Therefore, cell degradation is another important challenge to overcome on the passive DMFC technology. Despite the fact that the loss of efficiency during the fuel cell lifetime is unavoidable, its rate can be minimized through an understanding of the degradation and failure mechanisms. The durability of a fuel cell can be analyzed through the control of the current under the desired operating conditions and register the changes of the cell voltage over the time. In this test, a polarization curve is recorded at the beginning (BOL) and at the end (EOL) of the lifetime test and both results give the degradation rate, which is estimated using the voltage drop measured. Despite the usefulness of the durability tests, they do not allow to identify the different degradation and failure mechanisms that conduct to those losses, such as catalyst degradation and corrosion and membrane thinning and decomposition. Therefore, understanding the different degradation mechanism through the investigation of the state of the different fuel cell components, before and after the lifetime test is mandatory. It is necessary to use complementary electrochemical measurements, such as electrochemical impedance spectroscopy (EIS) [160, 165-168] and physical and chemical techniques, such as X-ray photoelectron spectroscopy (XPS), scanning electron microscopy (SEM), transmission electron microscopy (TEM) and Raman spectroscopy (RS), to identify the different specific phenomena that lead to the overall efficiency loss [160].

The fuel cell degradation can be permanent, which is irreversible, or temporary, which is reversible, meaning that performance losses during the cell operation can be recovered. Examples of temporary degradation are the cathode flooding and the carbon dioxide accumulation at the anode side hindering both oxygen a methanol transport [25, 26, 164]. To avoid the cathode flooding the cathode diffusion layers are usually PTFE treated to allow a hydrophobic surface. However, the degradation of the diffusion layer characterized by a decrease on its hydrophobic nature lead to a water accumulation on the cathode side, obstruction the oxygen diffusion to the catalyst sites [158, 162, 163, 167].

The permanent degradation has a chemical, thermal or mechanical nature and occur, due to the presence of foreign ions and contaminants, due to an increase of the cell temperature during cell operation and due to the compression during the cell assembly or MEA fabrication, respectively. The chemical degradation can cause changes on the chemical composition of the different fuel cell layers and membrane thickness, the thermal one lead to a decrease of the ionic conductivity and mechanical strength of the membrane and the mechanical degradation occur due to perforations, cracks and pinholes in the membrane [25, 26, 158, 160, 161, 162-167].

As already mentioned in DMFCs, Pt/Ru is used as anode catalyst to ensure an effective methanol oxidation and avoid the catalyst poisoning. However, at high current densities, Ru can cross the membrane towards the cathode, via the electro-osmotic water flux, reducing the CO poisoning tolerance of the anode catalyst and decreasing the anode electrochemical reaction rate [26, 160, 162, 164, 166-168].

The current collectors used in passive DMFCs are mainly made from stainless steel, but this material after a long-term operation suffer from corrosion, which increase the contact resistance between the current collector and the MEA and leads to the presence and accumulation of the corrosion products on membrane, catalyst and diffusion layers, poisoning them [25, 26, 161]. The presence of these compounds on the membrane lead to a decrease of the membrane conductivity and hinders the protons transport towards the cathode side [161].

ECONOMIC EVALUATION

As already referred, passive DMFCs are ideal candidates for replacing the conventional batteries in portable applications. However, one of the main barriers that impede its massive commercialization is its high costs. The passive DMFCs costs are higher than the Li-ion batteries ones due to the need of relatively high catalyst loadings of expensive catalytic materials, such as the noble metals Pt and Ru, the expensive Nafion membranes and the expensive metal coating (such as Au) on the current collector. As reported by a Korean company in 2007, the production cost of a 20 W DMFC for powering a laptop was estimated to be 297€, which is ten times higher than the cost of manufacturing a Li-ion battery with the same power output [169]. Moreover, the DOE target for the costs of portable DMFC systems is between 4.5€ W^{-1} and 6.2 € W^{-1} for systems of 100-250 W and 10-50 W in size, respectively.

In 2009, Rashidi et al. [3] studied the feasibility of using a passive DMFC to power portable devices for an operational period of 4 years through comparing these results with the ones obtained with a Li-ion battery. The results showed that, during the first year of operation, the conventional battery showed a lower cost than the passive DMFC. However, at the end of the 4 years of operation, the cost of the passive DMFC system were lower because during the operation time it was necessary to replace the battery more than once, which increase drastically its overall cost. Besides that, the CO_2 emissions during the operation of the DMFC were lower than the battery system leading to a lower carbon offset cost, proving that the DMFCs are a more environmentally friendly technology. Also, in 2009, Hashim et al. [170] developed a passive micro-DMFC and carried out a cost analysis, where concluded that the costs of using a micro-DMFC stack is equivalent to the costs of Li-ion batteries.

In the last years significant progress has been made in order to decrease the cost of passive DMFC systems through the development of new catalyst made from low cost materials, non-Pt catalyst, reduce the catalyst loading, new low cost membranes and current collectors with materials that have lower costs and higher corrosion resistance. In 2017,

Munjewar et al. [27] present a review paper concerning the material development for passive DMFC systems, among which were highlighted the development of electrolyte membranes, different catalysts and current collectors, based on the economic viability of the passive DMFCs. It was shown that the non-fluorinated membranes exhibited lower costs and methanol and Ru crossover rates, over the fluorinated (Nafion®) ones, but they were not competitive in terms of the fuel cell performance. The development of novel low-cost anode electro-catalyst with better kinetics and an optimized coating of the current collect in order to ensure and optimal balance between the corrosion resistance and its cost were identified as key factors towards the commercialization of passive DMFCs.

Conclusion

The fundamentals and challenges on passive DMFCs have been summarized and the recent modeling and experimental studies have been reviewed. An economic outlook of this technology was also presented. As was mentioned for practical applications of this technology cost effectiveness is essential. Moreover, the passive fuel cell systems must be made small and compact for portable applications, and each application has different power requirements.

In the last years, some efforts have been made on the development of passive DMFC systems with enhanced performances at an attractive cost. Some studies have been focused on the analysis and improvement of single parts of these systems and different materials, designs and operating conditions have been suggested. Although the results of these studies are useful, it is important to note that materials and conditions optimized for one type of fuel cell may be not always recommended for another. Therefore, when developing a passive DMFC, the materials and conditions need to be carefully chosen in order to achieve optimal performances.

A great challenge on the development of this type of fuel cell, in order to be used in real applications, is the development of cost-effective materials. To achieve that, different materials must be explored as current collectors, membrane, and diffusion and catalyst layers. Moreover, it is mandatory to reduce the actual catalyst loadings through the development of more active catalyst, with higher surface areas, by controlling its particle size distribution, morphology and crystallinity. Despite the work already performed, additional studies on cost-efficient electrode materials with optimized configurations are needed.

A key issue in DMFCs is the methanol crossover, since methanol diffuses through the membrane generating heat but no power. This can be limited if the cell operates with low methanol concentrations, but this significantly reduces the system energy density since large amounts of water are presented in the system and water will produce no power. In addition, this creates a higher water concentration gradient between the anode and cathode side, so more water crosses the membrane. The existence of large amounts of water on the cathode, due to crossover and water production on the cathode reaction, may lead to cathode flooding, with a consequent decrease on the cell performance. The solution is to use a high methanol concentration and use different membranes, diffusion and catalyst layers materials with different thicknesses.

Mathematical models appear as a useful tool to understand and predict the main transport phenomena and electrochemical processes helping in fuel cells optimization in terms of design and power output. In spite of the modelling work on passive DMFCs developed in the last years, a number of unsolved issues still demand for intensive research, such as the two-phase flow phenomena in both anode and the cathode sides. Such models will be helpful for the discovery of new cell designs and operation regimes for passive DMFC systems.

For commercialization of these devices, significant effort has to be put on the development of optimized systems with an optimum balance between its cost, efficiency, reliability and durability.

ACKNOWLEDGMENTS

B. A. Braz acknowledges the Ph.D. fellowship (BEX 12997/13-7) supported by CAPES Foundation, Ministry of Education of Brazil. V.B. Oliveira acknowledges the post-doctoral fellowship (SFRH/BDP/91993/

2012) supported by the Portuguese "Fundação para a Ciência e Tecnologia" (FCT), POPH/QREN and European Social Fund (ESF). POCI (FEDER) also supported this work via CEFT.

References

[1] Demirci, U. B. Direct liquid-feed fuel cells: Thermodynamic and environmental concerns. *J. Power Sources* 2007, *169* (2), 239–246.

[2] Sundmacher, K. Fuel Cell Engineering: Toward the Design of Efficient Electrochemical Power Plants. *Ind. Eng. Chem. Res.* 2010, *49* (21), 10159–10182.

[3] Rashidi, R.; Dincer, I.; Naterer, G. F.; Berg, P. Performance evaluation of direct methanol fuel cells for portable applications. *J. Power Sources* 2009, *187* (2), 509–516.

[4] Kamarudin, S. K.; Achmad, F.; Daud, W. R. W. Overview on the application of direct methanol fuel cell (DMFC) for portable electronic devices. *Int. J. Hydrogen Energy* 2009, *34* (16), 6902–6916.

[5] Zhao, X.; Yin, M.; Ma, L.; Liang, L.; Liu, C.; Liao, J.; Lu, T.; Xing, W. Recent advances in catalysts for direct methanol fuel cells. *Energy Environ. Sci.* 2011, *4* (8), 2736.

[6] Karim, N. a.; Kamarudin, S. K. An overview on non-platinum cathode catalysts for direct methanol fuel cell. *Appl. Energy* 2013, *103*, 212–220.

[7] Tiwari, J. N.; Tiwari, R. N.; Singh, G.; Kim, K. S. Recent progress in the development of anode and cathode catalysts for direct methanol fuel cells. *Nano Energy* 2013, *2* (5), 553–578.

[8] Singh, R. N.; Awasthi, R.; Sharma, C. S. Review: An Overview of Recent Development of Platinum- Based Cathode Materials for Direct Methanol Fuel Cells. *Int. J. Electrochem. Sci* 2014, *9*, 5607–5639.

[9] Kamarudin, S. K.; Hashim, N. Materials, morphologies and structures of MEAs in DMFCs. *Renew. Sustain. Energy Rev.* 2012, *16* (5), 2494–2515.

[10] Ahmad, H.; Kamarudin, S. K.; Hasran, U. a.; Daud, W. R. W. Overview of hybrid membranes for direct-methanol fuel-cell applications. *Int. J. Hydrogen Energy* 2010, *35* (5), 2160–2175.

[11] Lufrano, F.; Baglio, V.; Staiti, P.; Antonucci, V.; Arico', A. S. Performance analysis of polymer electrolyte membranes for direct methanol fuel cells. *J. Power Sources* 2013, *243*, 519–534.

[12] Zainoodin, A. M.; Kamarudin, S. K.; Daud, W. R. W. Electrode in direct methanol fuel cells. *Int. J. Hydrogen Energy* 2010, *35* (10), 4606–4621.

[13] Ahmed, M.; Dincer, I. A review on methanol crossover in direct methanol fuel cells : challenges and achievements. *Int. J. Energy Res.* 2011, *5*, 1213–1228.

[14] Zhao, T. S.; Yang, W. W.; Chen, R.; Wu, Q. X. Towards operating direct methanol fuel cells with highly concentrated fuel. *J. Power Sources* 2010, *195 (11)*, 3451–3462.

[15] Basri, S.; Kamarudin, S. K. Process system engineering in direct methanol fuel cell. *Int. J. Hydrogen Energy* 2011, *36* (10), 6219–6236.

[16] Faghri, A.; Li, X.; Bahrami, H. Recent advances in passive and semi-passive direct methanol fuel cells. *Int. J. Therm. Sci.* 2012, *62*, 12–18.

[17] Li, X.; Faghri, A. Review and advances of direct methanol fuel cells (DMFCs) part I: Design, fabrication, and testing with high concentration methanol solutions. *J. Power Sources* 2013, *226*, 223–240.

[18] Kamaruddin, M. Z. F.; Kamarudin, S. K.; Daud, W. R. W.; Masdar, M. S. An overview of fuel management in direct methanol fuel cells. *Renew. Sustain. Energy Rev.* 2013, *24*, 557–565.

[19] Yousefi, S.; Zohoor, M. Conceptual design and statistical overview on the design of a passive DMFC single cell. *Int. J. Hydrogen Energy* 2014, *39(11)*, 5972–5980.

[20] Kumar, P.; Dutta, K.; Das, S.; Kundu, P. P. An overview of unsolved deficiencies of direct methanol fuel cell technology: factors and parameters affecting its widespread use. *Int. J. Energy Res*. 2014, *38*, 1367–1390.

[21] Bahrami, H.; Faghri, A. Review and advances of direct methanol fuel cells: Part II: Modeling and numerical simulation. *J. Power Sources* 2013, *230*, 303–320.

[22] Ismail, A.; Kamarudin, S. K.; Daud, W. R. W.; Masdar, S.; Yosfiah, M. R. Mass and heat transport in direct methanol fuel cells. *J. Power Sources* 2011, *196* (23), 9847–9855.

[23] Zhao, T. S.; Xu, C.; Chen, R.; Yang, W. W. Mass transport phenomena in direct methanol fuel cells. *Prog. Energy Combust. Sci.* 2009, *35* (3), 275–292.

[24] Mallick, R. K.; Thombre, S. B.; Shrivastava, N. K. A critical review of the current collector for passive direct methanol fuel cells. *J. Power Sources* 2015, *285*, 510–529.

[25] Shrivastava, N. K.; Thombre, S. B.; Chadge, R. B. Liquid feed passive direct methanol fuel cell: challenges and recent advances. *Ionics (Kiel)*. 2016, *22 (1)*, 1–23.

[26] Munjewar, S. S.; Thombre, S. B.; Mallick, R. K. Approaches to overcome the barrier issues of passive direct methanol fuel cell - Review. *Renew. Sustain. Energy Rev.* 2017, *67*, 1087–1104.

[27] Munjewar, S. S.; Thombre, S. B.; Mallick, R. K. A comprehensive review on recent material development of passive direct methanol fuel cell. *Ionics (Kiel)*. 2017, *23 (1)*, 1–18.

[28] Chen, S.-L.; Lin, C.-T.; Chieng, C.-C.; Tseng, F.-G. Highly efficient CO_2 bubble removal on carbon nanotube supported nanocatalysts for direct methanol fuel cell. *J. Power Sources* 2010, *195 (6)*, 1640–1646.

[29] Yuan, W.; Tang, Y.; Yang, X.; Liu, B.; Wan, Z. Structural diversity and orientation dependence of a liquid-fed passive air-breathing direct methanol fuel cell. *Int. J. Hydrogen Energy* 2012, *37 (11)*, 9298–9313.

[30] Gholami, O.; Imen, S. J.; Shakeri, M. Effect of non-uniform parallel channel on performance of passive direct methanol fuel cell. *Int. J. Hydrogen Energy* 2013, *38 (8)*, 3395–3400.

[31] Yuan, W.; Zhou, B.; Hu, J.; Deng, J.; Zhang, Z.; Tang, Y. Passive direct methanol fuel cell using woven carbon fiber fabric as mass transfer control medium. *Int. J. Hydrogen Energy* 2015, *40 (5)*, 2326–2333.

[32] Falcão, D. S.; Pereira, J. P.; Pinto, A. M. F. R. Effect of stainless steel meshes on the performance of passive micro direct methanol fuel cells. *Int. J. Hydrogen Energy* 2016, *41 (31)*, 13859–13867.

[33] Zheng, W.; Suominen, A.; Lagercrantz, H.; Tuominen, A. A Two-Dimensional Model for CO2/Fuel Multiphase Flow on the Anode Side of a DMFC. *J. Fuel Cell Sci. Technol.* 2012, *9 (1)*, 11009.

[34] Liu, W.; Cai, W.; Liu, C.; Sun, S.; Xing, W. Magnetic coupled passive direct methanol fuel cell: Promoted CO_2 removal and enhanced catalyst utilization. *Fuel* 2015, *139*, 308–313.

[35] Zhou, Y.; Wang, X.; Guo, X.; Qiu, X.; Liu, L. A water collecting and recycling structure for silicon-based micro direct methanol fuel cells. *Int. J. Hydrogen Energy* 2012, *37 (1)*, 967–976.

[36] Xu, C.; Faghri, A. Effect of the Capillary Property of Porous Media on the Water Transport Characteristics in a Passive Liquid-Feed DMFC. *J. Fuel Cell Sci. Technol.* 2010, *7 (6)*, 61007.

[37] Li, X.; Faghri, A.; Xu, C. Water management of the DMFC passively fed with a high-concentration methanol solution. *Int. J. Hydrogen Energy* 2010, *35 (16)*, 8690–8698.

[38] Chen, M.; Chen, J.; Li, Y.; Huang, Q.; Zhang, H.; Xue, X.; Zou, Z.; Yang, H. Cathode Catalyst Layer with Stepwise Hydrophobicity Distribution for a Passive Direct Methanol Fuel Cell. *Energy & Fuels* 2012, *26 (2)*, 1178–1184.

[39] Yousefi, S.; Shakeri, M.; Sedighi, K. The effect of cell orientations and environmental conditions on the performance of a passive DMFC single cell. *Ionics (Kiel).* 2013, *19 (11)*, 1637–1647.

[40] Wang, Z.; Zhang, X.; Nie, L.; Zhang, Y.; Liu, X. Elimination of water flooding of cathode current collector of micro passive direct

methanol fuel cell by superhydrophilic surface treatment. *Appl. Energy* 2014, *126*, 107–112.

[41] Zhang, X.; Li, Y.; Chen, H.; Wang, Z.; Zeng, Z.; Cai, M.; Zhang, Y.; Liu, X. A water management system for metal-based micro passive direct methanol fuel cells. *J. Power Sources* 2015, *273*, 375–379.

[42] Bahrami, H.; Faghri, A. Water Management in a Passive DMFC Using Highly Concentrated Methanol Solution. *J. Fuel Cell Sci. Technol.* 2011, *8 (2)*, 21011.

[43] Xiao, B.; Bahrami, H.; Faghri, A. Analysis of heat and mass transport in a miniature passive and semi passive liquid-feed direct methanol fuel cell. *J. Power Sources* 2010, *195 (8)*, 2248–2259.

[44] Xu, C.; Faghri, A.; Li, X.; Ward, T. Methanol and water crossover in a passive liquid-feed direct methanol fuel cell. *Int. J. Hydrogen Energy* 2010, *35 (4)*, 1769–1777.

[45] Feng, L.; Zhang, J.; Cai, W.; Xing, W.; Liu, C. Single passive direct methanol fuel cell supplied with pure methanol. *J. Power Sources* 2011, *196 (5)*, 2750–2753.

[46] Park, Y.-C.; Kim, D.-H.; Lim, S.; Kim, S.-K.; Peck, D.-H.; Jung, D.-H. Design of a MEA with multi-layer electrodes for high concentration methanol DMFCs. *Int. J. Hydrogen Energy* 2012, *37 (5)*, 4717–4727.

[47] Yuan, W.; Tang, Y.; Yang, X.; Liu, B.; Wan, Z. Manufacture, characterization and application of porous metal-fiber sintered felt used as mass-transfer controlling medium for direct methanol fuel cells. *Trans. Nonferrous Met. Soc. China* 2013, *23 (7)*, 2085–2093.

[48] Yuan, W.; Tang, Y.; Yang, X.; Wan, Z. Toward using porous metal-fiber sintered plate as anodic methanol barrier in a passive direct methanol fuel cell. *Int. J. Hydrogen Energy* 2012, *37 (18)*, 13510–13521.

[49] Yuan, T.; Yang, J.; Wang, Y.; Ding, H.; Li, X.; Liu, L.; Yang, H. Anodic diffusion layer with graphene-carbon nanotubes composite material for passive direct methanol fuel cell. *Electrochim. Acta* 2014, *147*, 265–270.

[50] Huang, Q.; Jiang, J.; Chai, J.; Yuan, T.; Zhang, H.; Zou, Z.; Zhang, X.; Yang, H. Construction of porous anode by sacrificial template for a passive direct methanol fuel cell. *J. Power Sources* 2014, *262*, 213–218.

[51] Abdelkareem, M. A.; Kasem, E. T.; Nakagawa, N.; Abdelghani, E. A. M.; Elzatahry, A. A.; Khalil, K. A.; Barakat, N. A. M. Enhancement of the Passive Direct Methanol Fuel Cells Performance by Modification of the Cathode Microporous Layer Using Carbon Nanofibers. *Fuel Cells* 2014, *14 (4)*, 607–613.

[52] Wu, H.; Yuan, T.; Huang, Q.; Zhang, H.; Zou, Z.; Zheng, J.; Yang, H. Polypyrrole nanowire networks as anodic micro-porous layer for passive direct methanol fuel cells. *Electrochim. Acta* 2014, *141*, 1–5.

[53] Zhang, Y.; Xue, R.; Zhang, X.; Song, J.; Liu, X. RGO deposited in stainless steel fiber felt as mass transfer barrier layer for μ-DMFC. *Energy* 2015, *91*, 1081–1086.

[54] Oliveira, V. B.; Pereira, J. P.; Pinto, A. M. F. R. Effect of anode diffusion layer (GDL) on the performance of a passive direct methanol fuel cell (DMFC). *Int. J. Hydrogen Energy* 2016, *41* (42), 19455–19462.

[55] Shaffer, C. E.; Wang, C.-Y. Role of hydrophobic anode MPL in controlling water crossover in DMFC. *Electrochim. Acta* 2009, *54 (24)*, 5761–5769.

[56] Cao, J.; Chen, M.; Chen, J.; Wang, S.; Zou, Z.; Li, Z.; Akins, D. L.; Yang, H. Double microporous layer cathode for membrane electrode assembly of passive direct methanol fuel cells. *Int. J. Hydrogen Energy* 2010, *35 (10)*, 4622–4629.

[57] Shaffer, C. E.; Wang, C.-Y. High concentration methanol fuel cells: Design and theory. *J. Power Sources* 2010, *195 (13)*, 4185–4195.

[58] Wu, Q. X.; Zhao, T. S.; Chen, R.; Yang, W. W. Enhancement of water retention in the membrane electrode assembly for direct methanol fuel cells operating with neat methanol. *Int. J. Hydrogen Energy* 2010, *35 (19)*, 10547–10555.

[59] Peng, H.-C.; Chen, P.-H.; Chen, H.-W.; Chieng, C.-C.; Yeh, T.-K.; Pan, C.; Tseng, F.-G. Passive cathodic water/air management device for micro-direct methanol fuel cells. *J. Power Sources* 2010, *195 (21)*, 7349–7358.

[60] Wu, Q. X.; Zhao, T. S.; Yang, W. W. Effect of the cathode gas diffusion layer on the water transport behavior and the performance of passive direct methanol fuel cells operating with neat methanol. *Int. J. Heat Mass Transf.* 2011, *54 (5–6)*, 1132–1143.

[61] Zhang, J.; Feng, L.; Cai, W.; Liu, C.; Xing, W. The function of hydrophobic cathodic backing layers for high energy passive direct methanol fuel cell. *J. Power Sources* 2011, *196* (22), 9510–9515.

[62] Oliveira, V. B.; Falcão, D. S.; Rangel, C. M.; Pinto, A. M. F. R. Water management in a passive direct methanol fuel cell. *Int. J. Energy Res.* 2013, *37*, 991–1001.

[63] Deng, H.; Zhang, Y.; Zheng, X.; Li, Y.; Zhang, X.; Liu, X. A CNT (carbon nanotube) paper as cathode gas diffusion electrode for water management of passive □-DMFC (micro-direct methanol fuel cell) with highly concentrated methanol. *Energy* 2015, *82*, 236–241.

[64] Yan, X. H.; Zhao, T. S.; Zhao, G.; An, L.; Zhou, X. L. A hydrophilic-hydrophobic dual-layer microporous layer enabling the improved water management of direct methanol fuel cells operating with neat methanol. *J. Power Sources* 2015, *294*, 232–238.

[65] Wu, Q. X.; Zhao, T. S.; Chen, R.; Yang, W. W. A microfluidic-structured flow field for passive direct methanol fuel cells operating with highly concentrated fuels. *J. Micromechanics Microengineering* 2010, *20 (4)*, 45014.

[66] Cai, W.; Li, S.; Feng, L.; Zhang, J.; Song, D.; Xing, W.; Liu, C. Transient behavior analysis of a new designed passive direct methanol fuel cell fed with highly concentrated methanol. *J. Power Sources* 2011, *196 (8)*, 3781–3789.

[67] Yuan, W.; Tang, Y.; Wan, Z.; Pan, M. Operational characteristics of a passive air-breathing direct methanol fuel cell under various structural conditions. *Int. J. Hydrogen Energy* 2011, *36 (3)*, 2237–2249.

[68] Yousefi, S.; Ganji, D. D. Experimental investigation of a passive direct methanol fuel cell with 100 cm^2 active areas. *Electrochim. Acta* 2012, *85*, 693–699.

[69] Li, X.; Faghri, A. Development of a direct methanol fuel cell stack fed with pure methanol. *Int. J. Hydrogen Energy* 2012, *37 (19)*, 14549–14556.

[70] Yuan, W.; Tang, Y.; Yang, X. High-concentration operation of a passive air-breathing direct methanol fuel cell integrated with a porous methanol barrier. *Renew. Energy* 2013, *50*, 741–746.

[71] Guo, J.; Zhang, H.; Jiang, J.; Huang, Q.; Yuan, T.; Yang, H. Development of a self-adaptive direct methanol fuel cell fed with 20 M methanol. *Fuel Cells* 2013, *13 (6)*, 1018–1023.

[72] Yan, X. H.; Zhao, T. S.; An, L.; Zhao, G.; Zeng, L. A micro-porous current collector enabling passive direct methanol fuel cells to operate with highly concentrated fuel. *Electrochim. Acta* 2014, *139*, 7–12.

[73] Wu, H.; Zhang, H.; Chen, P.; Guo, J.; Yuan, T.; Zheng, J.; Yang, H. Integrated anode structure for passive direct methanol fuel cells with neat methanol operation. *J. Power Sources* 2014, *248*, 1264–1269.

[74] Wu, Q. X.; An, L.; Yan, X. H.; Zhao, T. S. Effects of design parameters on the performance of passive direct methanol fuel cells fed with concentrated fuel. *Electrochim. Acta* 2014, *133*, 8–15.

[75] Yuan, W.; Deng, J.; Zhang, Z.; Yang, X.; Tang, Y. Study on operational aspects of a passive direct methanol fuel cell incorporating an anodic methanol barrier. *Renew. Energy* 2014, *62*, 640–648.

[76] Abdelkareem, M. A.; Yoshitoshi, T.; Tsujiguchi, T.; Nakagawa, N. Vertical operation of passive direct methanol fuel cell employing a porous carbon plate. *J. Power Sources* 2010, *195*, 1821–1828.

[77] Yousefi, S.; Shakeri, M.; Sedighi, K. The effect of operating parameters on the perfomance of a passive DMFC single cell. *World Appl. Sci. J.* 2013, *22 (4)*, 516–522.

[78] Huang, J.; Ward, T.; Faghri, A. Optimizing the Anode Structure of a Passive Tubular-Shaped Direct Methanol Fuel Cell to Operate With High Concentration Methanol. *J. Fuel Cell Sci. Technol.* 2012, *9*, 51008.

[79] Liu, X.; Fang, S.; Ma, Z.; Zhang, Y. Structure design and implementation of the passive □-DMFC. *Micromachines* 2015, *6 (2)*, 230–238.

[80] Falcão, D. S.; Pereira, J. P.; Rangel, C. M.; Pinto, A. M. F. R. Development and performance analysis of a metallic passive micro-direct methanol fuel cell for portable applications. *Int. J. Hydrogen Energy* 2015, *40 (15)* 5408-5415.

[81] Tsai, L. D.; Chien, H. C.; Huang, W. H.; Huang, C. P.; Kang, C. Y.; Lin, J. N.; Chang, F. C. Novel bilayer composite membrane for passive direct methanol fuel cells with pure methanol. *Int. J. Electrochem. Sci.* 2013, *8*, 9704–9713.

[82] Thiam, H. S.; Daud, W. R. W.; Kamarudin, S. K.; Mohamad, a. B.; Kadhum, a. a. H.; Loh, K. S.; Majlan, E. H. Performance of direct methanol fuel cell with a palladium–silica nanofibre/Nafion composite membrane. *Energy Convers. Manag.* 2013, *75*, 718–726.

[83] Yuan, T.; Pu, L.; Huang, Q.; Zhang, H.; Li, X.; Yang, H. An effective methanol-blocking membrane modified with graphene oxide nanosheets for passive direct methanol fuel cells. *Electrochim. Acta* 2014, *117*, 393–397.

[84] He, Q.; Zheng, J.; Zhang, S. Preparation and characterization of high performance sulfonated poly(p-phenylene-co-aryl ether ketone) membranes for direct methanol fuel cells. *J. Power Sources* 2014, *260*, 317–325.

[85] Cui, M.; Zhang, Z.; Yuan, T.; Yang, H.; Wu, L.; Bakangura, E.; Xu, T. Proton-conducting membranes based on side-chain-type sulfonated poly(ether ketone/ether benzimidazole)s via one-pot condensation. *J. Memb. Sci.* 2014, *465*, 100–106.

[86] Zheng, J.; Wang, J.; Zhang, S.; Yuan, T.; Yang, H. Synthesis of novel cardo poly(arylene ether sulfone)s with bulky and rigid side

chains for direct methanol fuel cells. *J. Power Sources* 2014, *245*, 1005–1013.

[87] Zheng, J.; He, Q.; Gao, N.; Yuan, T.; Zhang, S.; Yang, H. Novel proton exchange membranes based on cardo poly(arylene ether sulfone/nitrile)s with perfluoroalkyl sulfonic acid moieties for passive direct methanol fuel cells. *J. Power Sources* 2014, *261*, 38–45.

[88] Pu, L.; Jiang, J.; Yuan, T.; Chai, J.; Zhang, H.; Zou, Z.; Li, X.-M.; Yang, H. Performance improvement of passive direct methanol fuel cells with surface-patterned Nafion® membranes. *Appl. Surf. Sci.* 2015, *327*, 205–212.

[89] Yuan, T.; Kang, Y.; Chen, J.; Du, C.; Qiao, Y.; Xue, X.; Zou, Z.; Yang, H. Enhanced performance of a passive direct methanol fuel cell with decreased Nafion aggregate size within the anode catalytic layer. *Int. J. Hydrogen Energy* 2011, *36 (16)*, 10000–10005.

[90] Zheng, J.; He, Q.; Liu, C.; Yuan, T.; Zhang, S.; Yang, H. Nafion-microporous organic polymer networks composite membranes. *J. Memb. Sci.* 2015, *476*, 571–579.

[91] Jiang, Z.; Jiang, Z.-J. Plasma-Polymerized Membranes with High Proton Conductivity for a Micro Semi-Passive Direct Methanol Fuel Cell. *Plasma Process. Polym.* 2016, *13 (1)*, 105–115.

[92] Zheng, J.; Bi, W.; Dong, X.; Zhu, J.; Mao, H.; Li, S.; Zhang, S. High performance tetra-sulfonated poly(p-phenylene-co-aryl ether ketone) membranes with microblock moieties for passive direct methanol fuel cells. *J. Memb. Sci.* 2016, *517*, 47–56.

[93] Yan, X. H.; Wu, R.; Xu, J. B.; Luo, Z.; Zhao, T. S. A monolayer graphene - Nafion sandwich membrane for direct methanol fuel cells. *J. Power Sources* 2016, *311*, 188–194.

[94] Gharibi, H.; Golmohammadi, F.; Kheirmand, M. Fabrication of MEA based on optimum amount of Co in PdxCo/C alloy nanoparticles as a new cathode for oxygen reduction reaction in passive direct methanol fuel cells. *Electrochim. Acta* 2013, *89*, 212–221.

[95] Gharibi, H.; Golmohammadi, F.; Kheirmand, M. Palladium/cobalt coated on multi-walled carbon nanotubes as an electro-catalyst for oxygen reduction reaction in passive direct methanol fuel cells. *Fuel Cells* 2013, *13 (6)*, 987–1004.

[96] Chen, P.; Wu, H.; Yuan, T.; Zou, Z.; Zhang, H.; Zheng, J.; Yang, H. Electronspun nanofiber network anode for a passive direct methanol fuel cell. *J. Power Sources* 2014, *255*, 70–75.

[97] Pu, L.; Zhang, H.; Yuan, T.; Zou, Z.; Zou, L.; Li, X.-M.; Yang, H. High performance platinum nanorod assemblies based double-layered cathode for passive direct methanol fuel cells. *J. Power Sources* 2015, *276*, 95–101.

[98] Ferreira E. C. L; Quadros F. M.; Souza J. P. I. Desenvolvimento de conjunto membrana-eletrodos para célula a combustível de metanol direto passiva. *Quim. Nova* 2010, *33 (6)*, 1313–1319.

[99] Asteazaran, M.; Cespedes, G.; Bengió, S.; Moreno, M. S.; Triaca, W. E.; Castro Luna, A. M. Research on methanol-tolerant catalysts for the oxygen reduction reaction. *J. Appl. Electrochem.* 2015, *45 (11)*, 1187–1193.

[100] Chai, J.; Zhou, Y.; Fan, J.; Jiang, J.; Yuan, T.; Zhang, H.; Zou, Z.; Qian, H.; Yang, H. Fabrication of nano-network structure anode by zinc oxide nanorods template for passive direct methanol fuel cells. *Int. J. Hydrogen Energy* 2015, *40 (20)*, 6647–6654.

[101] Amani, M.; Kazemeini, M.; Hamedanian, M.; Pahlavanzadeh, H.; Gharibi, H. Investigation of methanol oxidation on a highly active and stable Pt-Sn electrocatalyst supported on carbon-polyaniline composite for application in a passive direct methanol fuel cell. *Mater. Res. Bull.* 2015, *68*, 166–178.

[102] Wang, Y.; Cheng, Q.; Yuan, T.; Zhou, Y.; Zhang, H.; Zou, Z.; Fang, J.; Yang, H. Controllable fabrication of ordered Pt nanorod array as catalytic electrode for passive direct methanol fuel cells. *Chinese J. Catal.* 2016, *37 (7)*, 1089–1095.

[103] Golmohammadi, F.; Gharibi, H.; Sadeghi, S. Synthesis and electrochemical characterization of binary carbon supported Pd3Co

nanocatalyst for oxygen reduction reaction in direct methanol fuel cells. *Int. J. Hydrogen Energy* 2016, *41 (18)*, 7373–7387.

[104] Zainoodin, a. M.; Kamarudin, S. K.; Masdar, M. S.; Daud, W. R. W.; Mohamad, a. B.; Sahari, J. High power direct methanol fuel cell with a porous carbon nanofiber anode layer. *Appl. Energy* 2014, *113*, 946–954.

[105] Zainoodin, A. M.; Kamarudin, S. K.; Masdar, M. S.; Daud, W. R. W.; Mohamad, A. B.; Sahari, J. Investigation of MEA degradation in a passive direct methanol fuel cell under different modes of operation. *Appl. Energy* 2014, *135*, 364–372.

[106] Shrivastava, N. K.; Thombre, S. B.; Mallick, R. K. Effect of diffusion layer compression on passive DMFC performance. *Electrochim. Acta* 2014, *149*, 167–175.

[107] Cao, J.; Wang, L.; Song, L.; Xu, J.; Wang, H.; Chen, Z.; Huang, Q.; Yang, H. Novel cathodal diffusion layer with mesoporous carbon for the passive direct methanol fuel cell. *Electrochim. Acta* 2014, *118*, 163–168.

[108] Yuan, W.; Tang, Y.; Wang, Q.; Wan, Z. Dominance evaluation of structural factors in a passive air-breathing direct methanol fuel cell based on orthogonal array analysis. *Appl. Energy* 2011, *88 (5)*, 1671–1680.

[109] Ong, B. C.; Kamarudin, S. K.; Masdar, M. S.; Hasran, U. A. Applications of graphene nano-sheets as anode diffusion layers in passive direct methanol fuel cells (DMFC). *Int. J. Hydrogen Energy* 2017, *42 (14)* 9252–9261.

[110] Shrivastava, N. K.; Thombre, S. B.; Motghare, R. V. Wire mesh current collectors for passive direct methanol fuel cells. *J. Power Sources* 2014, *272*, 629–638.

[111] Calabriso, A.; Cedola, L.; Zotto, L. Del; Rispoli, F.; Santori, S. G. Performance investigation of Passive Direct Methanol Fuel Cell in different structural con fi gurations. *J. Clean. Prod.* 2015, *88*, 23–28.

[112] Tang, Y.; Yuan, W.; Pan, M.; Tang, B.; Li, Z.; Wan, Z. Effects of structural aspects on the performance of a passive air-breathing direct methanol fuel cell. *J. Power Sources* 2010, *195 (17)*, 5628–5636.

[113] Gholami, O.; Imen, S. J.; Shakeri, M. Effect of anode and cathode flow field geometry on passive direct methanol fuel cell performance. *Electrochim. Acta* 2015, *158*, 410–417.

[114] Mallick, R. K.; Thombre, S. B.; Motghare, R. V.; Chillawar, R. R. Analysis of the clamping effects on the passive direct methanol fuel cell performance using electrochemical impedance spectroscopy. *Electrochim. Acta* 2016, *215*, 150–161.

[115] Hashemi, R.; Yousefi, S.; Faraji, M. Experimental studying of the effect of active area on the performance of passive direct methanol fuel cell. *Ionics (Kiel).* 2015, *21 (10)*, 2851–2862.

[116] Zheng, W.; Suominen, A.; Kankaanranta, J.; Tuominen, A. A new structure of a passive direct methanol fuel cell. *Chem. Eng. Sci.* 2012, *76*, 188–191.

[117] Ward, T.; Li, X.; Faghri, A. Performance characteristics of a novel tubular-shaped passive direct methanol fuel cell. *J. Power Sources* 2011, *196 (15)*, 6264–6273.

[118] Tsujiguchi, T.; Abdelkareem, M. A.; Kudo, T.; Nakagawa, N.; Shimizu, T.; Matsuda, M. Development of a passive direct methanol fuel cell stack for high methanol concentration. *J. Power Sources* 2010, *195 (18)*, 5975–5979.

[119] Achmad, F.; Kamarudin, S. K.; Daud, W. R. W.; Majlan, E. H. Passive direct methanol fuel cells for portable electronic devices. *Appl. Energy* 2011, *88 (5)*, 1681–1689.

[120] Feng, L.; Cai, W.; Li, C.; Zhang, J.; Liu, C.; Xing, W. Fabrication and performance evaluation for a novel small planar passive direct methanol fuel cell stack. *Fuel* 2012, *94*, 401–408.

[121] Baglio, V.; Stassi, a.; Matera, F. V.; Kim, H.; Antonucci, V.; Aricò, a. S. AC-impedance investigation of different MEA configurations for passive-mode DMFC mini-stack applications. *Fuel Cells* 2010, *10 (1)*, 124–131.

[122] Wang, L.; He, M.; Hu, Y.; Zhang, Y.; Liu, X.; Wang, G. A “4-cell” modular passive DMFC (direct methanol fuel cell) stack for portable applications. *Energy* 2015, *82*, 229–235.

[123] Yuan, W.; Zhang, X.; Zhang, S.; Hu, J.; Li, Z.; Tang, Y. Lightweight current collector based on printed-circuit-board technology and its structural effects on the passive air-breathing direct methanol fuel cell. *Renew. Energy* 2015, *81*, 664–670.

[124] Barbera, O.; Stassi, A.; Sebastian, D.; Bonde, J. L.; Giacoppo, G.; D'Urso, C.; Baglio, V.; Aricò, A. S. Simple and functional direct methanol fuel cell stack designs for application in portable and auxiliary power units. *Int. J. Hydrogen Energy* 2016, *41 (28)*, 12320–12329.

[125] Masdar, M. S.; Zainoodin, A. M.; Rosli, M. I.; Kamarudin, S. K.; Daud, W. R. W. Performance and stability of single and 6-cell stack passive direct methanol fuel cell (DMFC) for long-term operation. *Int. J. Hydrogen Energy* 2015, *42 (14)* 1–13.

[126] Oliveira, V. B.; Rangel, C. M.; Pinto, A. M. F. R. One-dimensional and non-isothermal model for a passive DMFC. *J. Power Sources* 2011, *196* (21), 8973–8982.

[127] Guarnieri, M.; Di Noto, V.; Moro, F. A Dynamic Circuit Model of a Small Direct Methanol Fuel Cell for Portable Electronic Devices. *IEEE Trans. Ind. Electron.* 2010, *57* (6), 1865–1873.

[128] Bahrami, H.; Faghri, A. Exergy analysis of a passive direct methanol fuel cell. *J. Power Sources* 2011, *196* (3), 1191–1204.

[129] Rosenthal, N. S.; Vilekar, S. a.; Datta, R. A comprehensive yet comprehensible analytical model for the direct methanol fuel cell. *J. Power Sources* 2012, *206*, 129–143.

[130] Shrivastava, N. K.; Thombre, S. B.; Wasewar, K. L. Nonisothermal Mathematical Model for Performance Evaluation of Passive Direct Methanol Fuel Cells. *J. Energy Eng.* 2013, *139* (4), 266–274.

[131] Ko, J.; Chippar, P.; Ju, H. A one-dimensional, two-phase model for direct methanol fuel cells – Part I: Model development and parametric study. *Energy* 2010, *35* (5), 2149–2159.

[132] Yang, W. W.; Zhao, T. S.; Wu, Q. X. Modeling of a passive DMFC operating with neat methanol. *Int. J. Hydrogen Energy* 2011, *36 (11)*, 6899–6913.

[133] Cai, W.; Li, S.; Yan, L.; Feng, L.; Zhang, J.; Liang, L.; Xing, W.; Liu, C. Design and simulation of a liquid electrolyte passive direct methanol fuel cell with low methanol crossover. *J. Power Sources* 2011, *196 (18)*, 7616–7626.

[134] Cai, W.; Li, S.; Li, C.; Liang, L.; Xing, W.; Liu, C. A model based thermal management of DMFC stack considering the double-phase flow in the anode. *Chem. Eng. Sci.* 2013, *93*, 110–123.

[135] Yin, K. M. One-dimensional steady state algebraic model on the passive direct methanol fuel cell with consideration of the intermediate liquid electrolyte. *J. Power Sources* 2015, *282*, 368–377.

[136] Shrivastava, N. K.; Thombre, S. B.; Wasewar, K. L. A real-time simulating non-isothermal mathematical model for the passive feed direct methanol fuel cell. *Int. J. Green Energy* 2016, *13 (2)*, 213–228.

[137] Basri, S.; Kamarudin, S. K.; Daud, W. R. W.; Yaakub, Z.; Ahmad, M. M.; Hashim, N.; Hasran, U. a. Unsteady-state modelling for a passive liquid-feed DMFC. *Int. J. Hydrogen Energy* 2009, *34 (14)*, 5759–5769.

[138] Basri, S.; Kamarudin, S. K.; Daud, W. R. W.; Ahmad, M. M. Non-linear optimization of passive direct methanol fuel cell (DMFC). *Int. J. Hydrogen Energy* 2010, *35 (4)*, 1759–1768.

[139] Ye, D.; Zhu, X.; Liao, Q.; Li, J.; Fu, Q. Two-dimensional two-phase mass transport model for methanol and water crossover in air-breathing direct methanol fuel cells. *J. Power Sources* 2009, *192 (2)*, 502–514.

[140] Xu, C.; Faghri, A. Water transport characteristics in a passive liquid-feed DMFC. *Int. J. Heat Mass Transf.* 2010, *53 (9–10)*, 1951–1966.

[141] Bahrami, H.; Faghri, A. Transport phenomena in a semi-passive direct methanol fuel cell. *Int. J. Heat Mass Transf.* 2010, *53 (11–12)*, 2563–2578.

[142] Li, X.; Faghri, A. Local entropy generation analysis on passive high-concentration DMFCs (direct methanol fuel cell) with different cell structures. *Energy* 2011, *36 (1)*, 403–414.

[143] Ward, T.; Xu, C.; Faghri, A. Performance and design analysis of tubular-shaped passive direct methanol fuel cells. *Int. J. Hydrogen Energy* 2011, *36 (15)*, 9216–9230.

[144] Jewett, G.; Faghri, A.; Xiao, B. Optimization of water and air management systems for a passive direct methanol fuel cell. *Int. J. Heat Mass Transf.* 2009, *52 (15–16)*, 3564–3575.

[145] Gerteisen, D. Transient and steady-state analysis of catalyst poisoning and mixed potential formation in direct methanol fuel cells. *J. Power Sources* 2010, *195 (19)*, 6719–6731.

[146] Wang, L.; Zhang, Y.; An, Z.; Huang, S.; Zhou, Z.; Liu, X. Non-isothermal modeling of a small passive direct methanol fuel cell in vertical operation with anode natural convection effect. *Energy* 2013, *58*, 283–295.

[147] Li, X. Y.; Yang, W. W.; He, Y. L.; Zhao, T. S.; Qu, Z. G. Effect of anode micro-porous layer on species crossover through the membrane of the liquid-feed direct methanol fuel cells. *Appl. Therm. Eng.* 2012, *48*, 392–401.

[148] Miao, Z.; Xu, J.-L.; He, Y.-L. Modeling of the Transport Phenomena in Passive Direct Methanol Fuel Cells Using a Two-Phase Anisotropic Model. *Adv. Mech. Eng.* 2015, *15*, 1–15.

[149] Guo, H.; Chen, Y. P.; Xue, Y. Q.; Ye, F.; Ma, C. F. Three-dimensional transient modeling and analysis of two-phase mass transfer in air-breathing cathode of a fuel cell. *Int. J. Hydrogen Energy* 2013, *38 (25)*, 11028–11037.

[150] Guo, T.; Sun, J.; Deng, H.; Jiao, K.; Huang, X. Transient analysis of passive direct methanol fuel cells with different operation and design parameters. *Int. J. Hydrogen Energy* 2015, *40 (43)*, 14978–14995.

[151] Yuan, Z.; Yang, J.; Ye, N.; Li, Z.; Sun, Y.; Shen, H. Analysis of the capillary-force-based □DMFC (micro direct methanol fuel cell) supplied with pure methanol. *Energy* 2015, *89*, 858–863.

[152] Ting, G.; Sun, J.; Deng, H.; Xie, X.; Jiao, K.; Huang, X. Investigation of cell orientation effect on transient operation of passive direct methanol fuel cells. *Int. J. Hydrogen Energy* 2016, *41 (15)*, 6493–6507.

[153] Zago, M.; Casalegno, A.; Bresciani, F.; Marchesi, R. Effect of anode MPL on water and methanol transport in DMFC: Experimental and modeling analyses. *Int. J. Hydrogen Energy* 2014, *39 (36)*, 21620–21630.

[154] Sun, J.; Guo, T.; Deng, H.; Jiao, K.; Huang, X. Effect of electrode variable contact angle on the performance and transport characteristics of passive direct methanol fuel cells. *Int. J. Hydrogen Energy* 2015, *40 (33)*, 10568–10587.

[155] García-Salaberri, P. A.; Vera, M. On the effects of assembly compression on the performance of liquid-feed DMFCs under methanol-limiting conditions: A 2D numerical study. *J. Power Sources* 2015, *285*, 543–558.

[156] Xue, Y. Q.; Guo, H.; Shang, H. H.; Ye, F.; Ma, C. F. Simulation of mass transfer in a passive direct methanol fuel cell cathode with perforated current collector. *Energy* 2015, *81*, 501–510.

[157] Kamaruddin, M. Z. F.; Kamarudin, S. K.; Masdar, M. S.; Daud, W. R. W. Investigating design parameter effects on the methanol flux in the passive storage of a direct methanol fuel cell. *Int. J. Hydrogen Energy* 2015, *40 (35)*, 11931–11942.

[158] Mehmood, A.; Scibioh, M. A.; Prabhuram, J.; An, M. G.; Ha, H. Y. A review on durability issues and restoration techniques in long-term operations of direct methanol fuel cells. *J. Power Sources* 2015, *297*, 224–241.

[159] Knights, S. D.; Colbow, K. M.; St-Pierre, J.; Wilkinson, D. P. Aging mechanisms and lifetime of PEFC and DMFC. *J. Power Sources* 2004, *127 (1–2)*, 127–134.

[160] Chen, W.; Sun, G.; Guo, J.; Zhao, X.; Yan, S.; Tian, J.; Tang, S.; Zhou, Z.; Xin, Q. Test on the degradation of direct methanol fuel cell. 2006, *51*, 2391–2399.

[161] Guo, J.-W.; Xie, X.-F.; Wang, J.-H.; Shang, Y.-M. Effect of current collector corrosion made from printed circuit board (PCB) on the degradation of self-breathing direct methanol fuel cell stack. *Electrochim. Acta* 2008, *53 (7)*, 3056–3064.

[162] Cha, H.-C.; Chen, C.-Y.; Shiu, J.-Y. Investigation on the durability of direct methanol fuel cells. *J. Power Sources* 2009, *192 (2)*, 451–456.

[163] Bresciani, F.; Rabissi, C.; Zago, M.; Marchesi, R.; Casalegno, A. On the effect of gas diffusion layers hydrophobicity on direct methanol fuel cell performance and degradation. *J. Power Sources* 2015, *273*, 680–687.

[164] Zainoodin, a. M.; Kamarudin, S. K.; Masdar, M. S.; Daud, W. R. W.; Mohamad, a. B.; Sahari, J. Investigation of MEA degradation in a passive direct methanol fuel cell under different modes of operation. *Appl. Energy* 2014, *135*, 364–372.

[165] Jeon, M. K.; Won, J. Y.; Oh, K. S.; Lee, K. R.; Woo, S. I. Performance degradation study of a direct methanol fuel cell by electrochemical impedance spectroscopy. *Electrochim. Acta* 2007, *53 (2)*, 447–452.

[166] Lai, C.-M.; Lin, J.-C.; Hsueh, K.-L.; Hwang, C.-P.; Tsay, K.-C.; Tsai, L.-D.; Peng, Y.-M. On the Accelerating Degradation of DMFC at Highly Anodic Potential. *J. Electrochem. Soc.* 2008, *155 (8)*, B843–B851.

[167] Chen, C. Y.; Cha, H. C. Strategy to optimize cathode operating conditions to improve the durability of a Direct Methanol Fuel Cell. *J. Power Sources* 2012, *200*, 21–28.

[168] Escudero-Cid, R.; Pérez-Flores, J. C.; Fatás, E.; Ocón, P. Degradation of DMFC using a New Long-Term Stability Cycle. *Int. J. Green Energy* 2015, *12*, 641–653.

[169] Joghee, P.; Malik, J.; Pylypenko, S.; & O'Hayre, R. A review on direct methanol fuel cells – In the perspective of energy and sustainability. *MRS Energy & Sustainability* 2015, *2*, 1-31.

[170] Hashim, N.; Kamarudin, S. K.; Daud, W. R. W. Design, fabrication and testing of a PMMA-based passive single-cell and a multi-cell stack micro-DMFC. *Int. J. Hydrogen Energy* 2009, *34 (19)*, 8263–8269.

INDEX

C

D

E

F

G

H

I

K

L

M

N

O

P

R

S

T

U

V

W

X